HOME REPAIR AND IMPROVEMENT

FINISH CARPENTRY

TIME® LIFE BOOKS

Other Publications:

DO IT YOURSELF
Custom Woodworking
Golf Digest Total Golf
How to Fix It
The Time-Life Complete Gardener
The Art of Woodworking

COOKING
Weight Watchers® Smart Choice Recipe Collection
Great Taste~Low Fat
Williams-Sonoma Kitchen Library

HISTORY
Our American Century
World War II
What Life Was Like
The American Story
Voices of the Civil War
The American Indians
Lost Civilizations
Mysteries of the Unknown
Time Frame
The Civil War
Cultural Atlas

TIME-LIFE KIDS
Student Library
Library of First Questions and Answers
A Child's First Library of Learning
I Love Math
Nature Company Discoveries
Understanding Science & Nature

SCIENCE/NATURE
Voyage Through the Universe

*For information on and a full description
of any of the Time-Life Books series listed above,
please call 1-800-621-7026 or write:*
Reader Information
Time-Life Customer Service
P.O. Box C-32068
Richmond, Virginia 23261-2068

10 9 8 7 6 5 4 3 2 1

HOME REPAIR AND IMPROVEMENT

FINISH CARPENTRY

BY THE EDITORS OF TIME-LIFE BOOKS, ALEXANDRIA, VIRGINIA

The Consultants

Karl Marcuse has been working as a self-employed carpenter and contractor for almost 30 years. His work has taken him to many regions of the world, from the high Arctic to the Far East. Specializing in house restoration, Mr. Marcuse lives in his fully restored century-old home in Montreal, Quebec.

CONTENTS

1 A GALLERY OF WALL TREATMENTS — 6

- Decorative Wood Moldings — 8
- Running Baseboard — 12
- Installing Crown Molding — 20
- Embellishing Walls with Rails — 24
- Tongue-and-Groove Wainscoting — 26
- Frame-and-Panel Wainscoting — 31
- Creating a Fireplace Surround — 36

2 WINDOWS AND TRIM — 42

- Installing a Window — 44
- Picture-Frame Casing — 46
- Stool-and-Apron Trim — 53

3 HANGING DOORS — 58

- Building a Frame-and-Panel Door — 60
- Installing Jambs — 65
- Hanging the Door — 69
- Installing a Lockset — 73
- Finishing Up with Trim — 77
- A Formal Entryway — 80

4 CRAFTING A STAIRCASE — 90

- Planning and Design — 92
- Constructing a Solid Landing — 98
- Fitting the Stringers — 101
- Cutting and Setting Newel Posts — 108
- Adding Skirts, Risers, and Treads — 114
- Putting in Handrails and Balusters — 120

Index — 126

Acknowledgments — 128

Picture Credits — 128

1

A Gallery of Wall Treatments

Whether simple or ornate, wood moldings define the character of a room. For such applications as baseboard, chair rail, and crown molding, trim can be purchased or fashioned in a wide range of profiles to complement the architectural style of any home; and wall paneling, or wainscoting, can bring a rich, warm look to an otherwise ordinary room.

Decorative Wood Moldings — 8

Shaping Trim on a Table Saw
Cutting Profiles on a Router Table

Running Baseboard — 12

Planning the Installation
Mitering Inside Corners
Coping a Joint
Fitting Outside Corners
Splicing Trim For a Long Wall
Adding Shoe Molding

Installing Crown Molding — 20

Mitering at Outside Corners
Coping Inside Corners

Embellishing Walls with Rails — 24

Fastening Chair Rail

Tongue-and-Groove Wainscoting — 26

Fitting and Nailing the Boards

Frame-and-Panel Wainscoting — 31

Attaching the Plywood Base
Building Up the Surface

Creating a Fireplace Surround — 36

Assembling the Pieces
Installing the Unit
Adding the Trim

Coping a piece of molding →

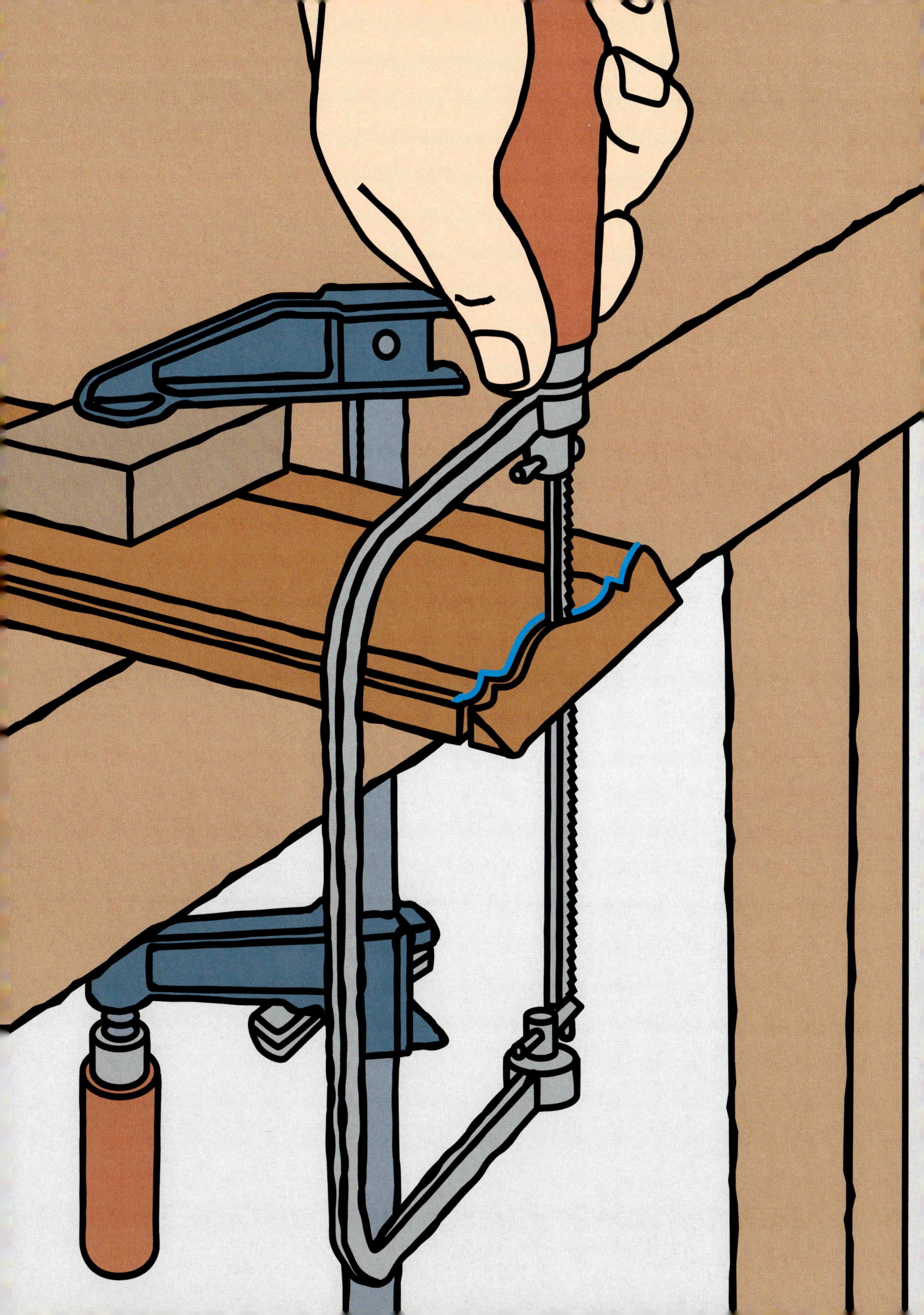

Decorative Wood Moldings

Wood molding is available in many styles at lumberyards and home centers, and is made for a range of purposes. If you are planning to furnish a room with new trim, you can probably find ready-made molding appropriate for the job. But when you are adding or replacing trim in a room with existing molding, you may want to fashion your own pieces to guarantee an exact match.

Buying Trim: Select molding whose size is appropriate for the proportions of the room in which you plan to install it. When choosing baseboard for a modern home with 8-foot-high ceilings, for example, resist the temptation to buy old-fashioned 10-inch stock.

If you plan to apply a clear finish to the trim, purchase solid wood. When you will be painting the molding, you can choose less expensive stock made of finger-jointed wood or medium-density fiberboard (MDF). Store the pieces in the room for a few weeks before installation to acclimatize them.

Shaping Molding: Each type of trim—whether it be baseboard, chair rail, crown molding, or other—has its own range of profiles. The best tools for forming these shapes are a table saw *(page 10)* or router *(page 11)*. Combining cuts with different blades or bits can create almost any profile you want.

Finishing the Pieces: Apply a primer before you install any trim that you plan to paint; some wood species also require a wood filler before priming *(opposite)*. Set nails and cover them with spackling compound, then plug any gaps with paintable caulk, sand the pieces, and apply a gloss or semigloss paint.

If you will be coating the trim with a clear finish, avoid leaving any gaps between the pieces as you install them, as you cannot use caulk. Set the nails and fill the holes with wood putty, then sand the wood, apply a stain if desired, then brush on polyurethane varnish.

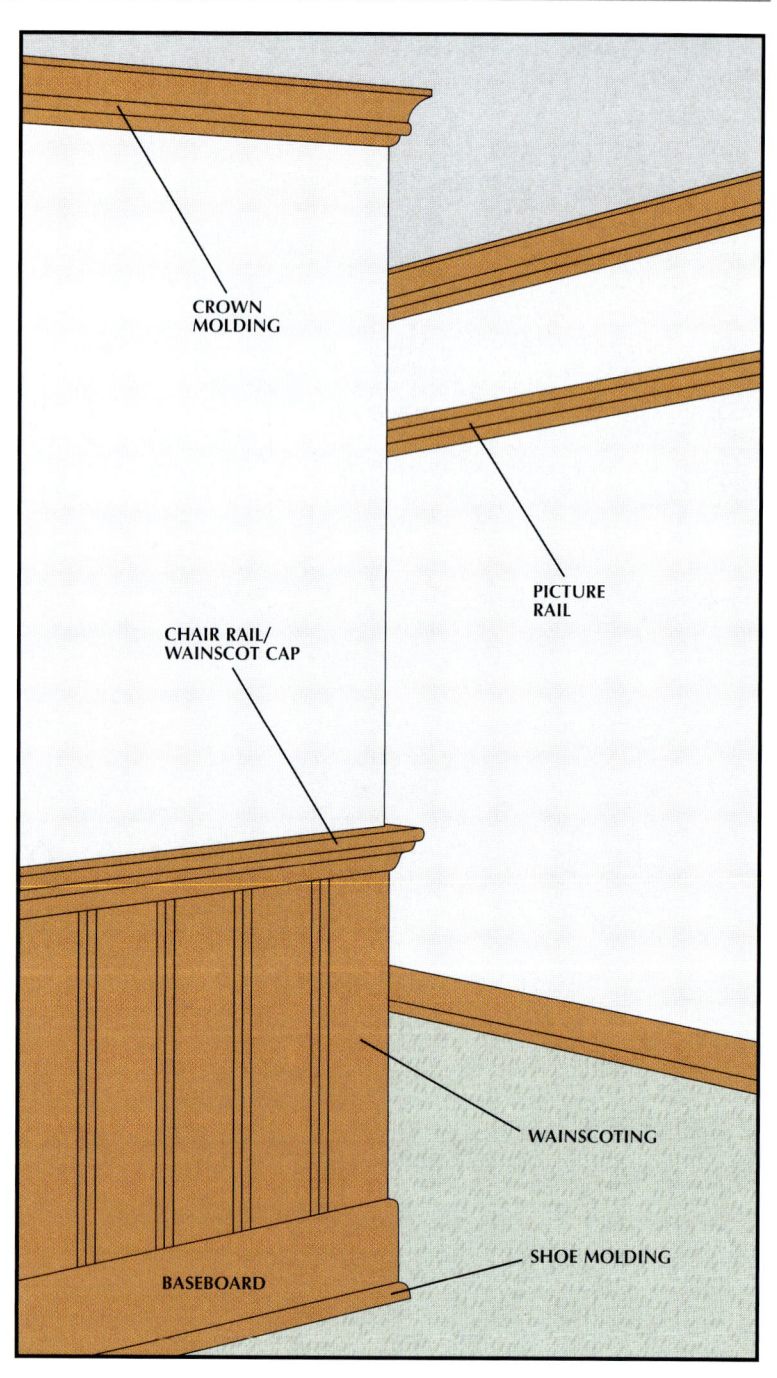

Molding types.
The drawing above includes every type of molding likely to be found in a home. Baseboard and shoe molding hide the gap between the floor and walls, while crown molding highlights the transition from walls to ceiling. Wainscoting covers the bottom third of the wall and gives a room a formal look. Originally intended to protect walls from chair backs, chair rail can serve to cap wainscoting or can be used alone, often with wallpaper below it. Picture rail, often found in older homes, is placed a couple of feet below the ceiling, and was designed for hanging pictures without marring the wall.

CHARACTERISTICS OF DIFFERENT WOOD SPECIES

Wood Species	Hardness	Workability	Dimensional Stability	Finishing
Ash	Hard	Fair	Fair	Accepts stains well; requires heavy filler for painting.
Beech	Very hard	Fair	Poor	Accepts stains well; requires thin filler for painting.
Birch	Hard	Good	Poor	Accepts finishes well.
Cedar, Western red	Soft	Good	Good	If staining, use oil stain.
Cherry	Medium	Good	Good	Accepts stains well; not suitable for painting.
Douglas-fir	Soft	Good	Good	If staining, use oil stain.
Maple, hard	Very hard	Difficult	Poor	Accepts finishes well.
Maple, soft	Medium	Fair	Poor	Accepts finishes well.
Oak, red	Hard	Good	Fair	Accepts stains well; requires heavy filler for painting.
Oak, white	Hard	Good	Poor	Accepts stains well; requires heavy filler for painting.
Pine, white	Soft	Good	Good	Accepts finishes well.
Pine, yellow	Medium	Fair	Fair	Accepts finishes well.
Redwood	Soft	Good	Good	If staining, use oil stain.
Walnut	Hard	Good	Good	Accepts stains well; not suitable for painting.

Selecting wood for trim.

In choosing a wood species for trim work, consider its color and grain pattern, as well as the other characteristics included in the chart above. Hardwoods are more difficult to nail, as they require pilot holes; however, they will also resist denting and other damage. Some woods are more workable than others; that is, they are easier to saw and shape. Dimensional stability refers to the amount the wood will shrink or expand with changes in humidity; joints in wood that is not dimensionally stable will have a tendency to open up with time. Wood species also vary in the way they take finishes—for instance, if a stain is to be applied, an oil-base product is recommended for certain woods, while other wood species require a wood filler before they can be painted.

SHAPING TRIM ON A TABLE SAW

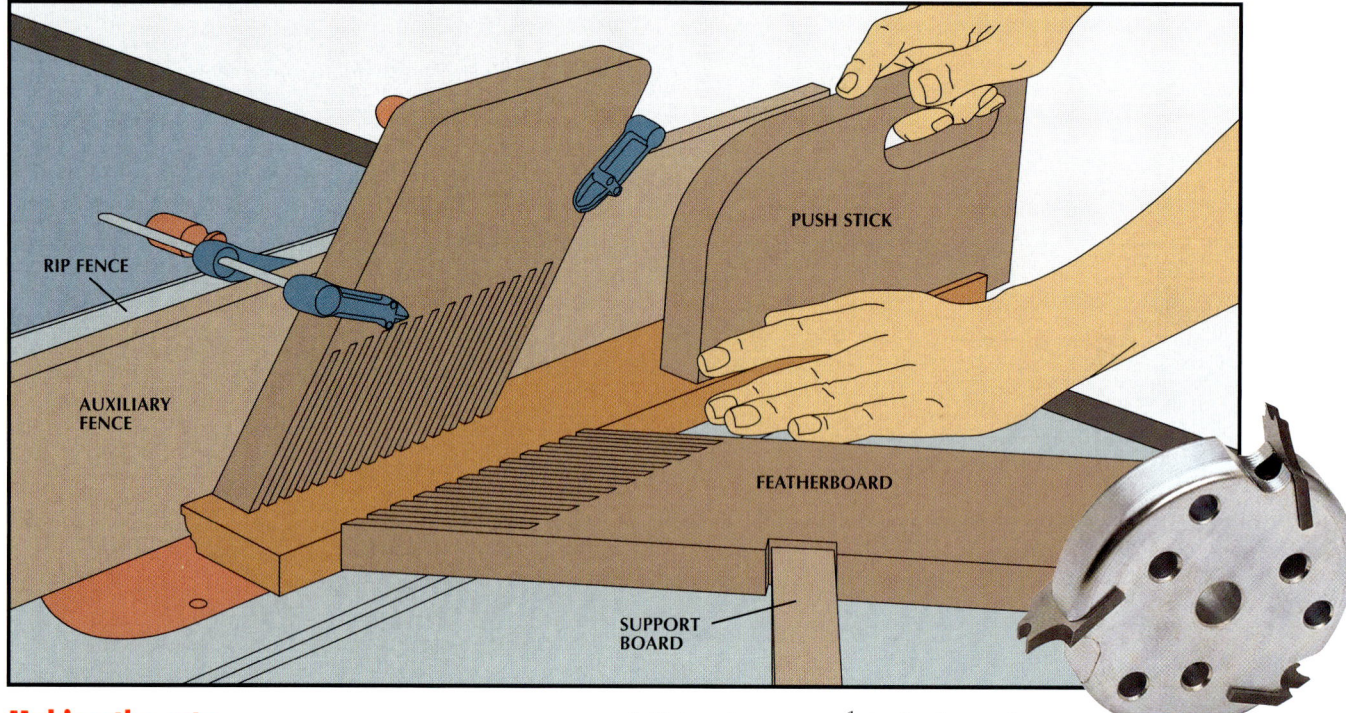

Making the cuts.
- Install a molding cutter head fitted with the knives that will create the profile *(photograph)*.
- Remove the blade guard from the saw, then clamp an auxiliary wood fence to the rip fence and position it over the cutter head. Keeping the metal fence clear of the knives, cut a clearance notch in the auxiliary fence by gradually raising the blade. Turn off the saw and position the fence for the desired profile.
- With the workpiece against the fence, clamp a featherboard *(box, below)* to the saw table, braced by a support board, so it presses the piece against the fence. Clamp a second featherboard vertically to the fence to hold the workpiece down on the saw table.
- Raise the cutters $\frac{1}{8}$ inch above the table and slowly feed the workpiece into them, pressing the stock against the fence with your free hand. Keep your hands well away from the blade and finish the cut with a push stick *(above)*.
- Make as many passes as necessary, raising the cutter head $\frac{1}{8}$ inch at a time until you achieve the desired depth of cut. For the final pass, raise the cutter head very slightly to produce a smooth finish.
- If desired, repeat the cut on the other edge of the board, or install a combination blade in the saw and cut the piece to the desired width.

A FEATHERBOARD FOR STRAIGHT CUTS

A featherboard keeps a workpiece firmly against the fence of a saw; its springy wood fingers allow the piece to move in only one direction—toward the blade. You can make a featherboard from a hardwood board that is $\frac{3}{4}$ inch thick, 6 inches wide, and about 16 inches long. Miter one end at 60 degrees, then cut a series of 5-inch-long kerfs at the mitered end, spacing them $\frac{1}{8}$ inch apart. To brace the featherboard against the saw table, cut a notch in the middle of the long edge to accommodate a $\frac{1}{2}$- by 1-inch support board that can be clamped to the table. You can also buy featherboards *(photograph)*; the model shown here has a metal bar that fits in the miter slot of the table saw; turning the knob tightens it in place and lowers the hold-down bracket against the workpiece.

CUTTING PROFILES ON A ROUTER TABLE

Fashioning the molding.

◆ Install a molding bit in the router and mount the tool in a table. Adjust the fence for a cut about $\frac{1}{8}$ inch wide.
◆ Clamp a featherboard *(opposite)* to the table in line with the bit and raised with a wood scrap to support the middle of the workpiece.
◆ Turn on the router and slowly feed the workpiece into the bit while holding it flush against the fence *(above, left)*. (In this illustration, the blade guard has been removed for the sake of clarity.) Finish the pass with a push stick *(above, right)*.
◆ Make as many passes as necessary to create the desired profile, increasing the cutting width by $\frac{1}{8}$ inch at a time.

TRICKS OF THE TRADE

Multiple Profiles with a Single Bit

You can shape almost any profile in wood by making a series of cuts with different router bits, but even a single bit can create several profiles, depending on how much of the bit protrudes above the table *(below)*. When experimenting with multiple cutting depths, make test passes on scrap wood first and check the profiles.

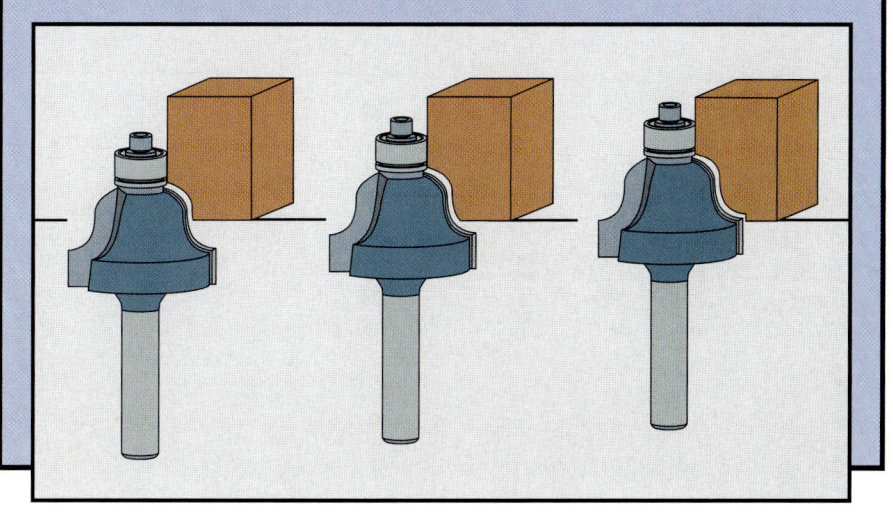

Running Baseboard

Where the walls of a room meet the floor, baseboard molding serves to cover gaps between the two surfaces. It may be designed as single sections or built up from two or more pieces *(below)*.

Planning the Layout: Lengths of baseboard are joined at corners and are spliced together on long walls. Corners and splices call for specific types of joints *(opposite, top)*. Plan cuts for joints to make the most efficient use of the wood and to hide imperfect pieces in closets or hallways. Also plan the installation sequence so that joints can be located as inconspicuously as possible *(opposite, bottom)*.

Installation: Baseboard is usually put in after the door and window casing and the finish floor are installed.

You can use a standard hammer to drive the nails, but a power finish nailer will speed the job *(page 14)*. Nail as close to the ends of pieces as possible without splitting the wood—$\frac{1}{2}$ inch for softwoods and 1 inch for hardwoods. With hardwood, drill pilot holes first if you are driving the nails with a hammer. Sink the nails with a nail set and fill the holes with wood putty or spackling compound.

TOOLS
Electronic stud finder
Power or manual miter saw
Electric drill
Hammer
Nail set
Clamps
Coping saw
Round file
Carpenter's square
Adjustable T-bevel
Utility knife

MATERIALS
Baseboard
Shoe molding
Finishing nails (2")
Construction adhesive
Wood glue

SAFETY TIPS
Protect your eyes with goggles when nailing or when operating a power tool.

Baseboard styles.
Baseboard ranges in style from single pieces to those built up from more than one element. The most common type of built-up baseboard incorporates shoe molding—usually either cove or quarter-round. A more elaborate style includes a cap as well as the shoe. Single-piece baseboard is easiest to install but the built-up type is better for an irregular wall, since the narrow shoe or cap molding will conform more easily to bumps or hollows.

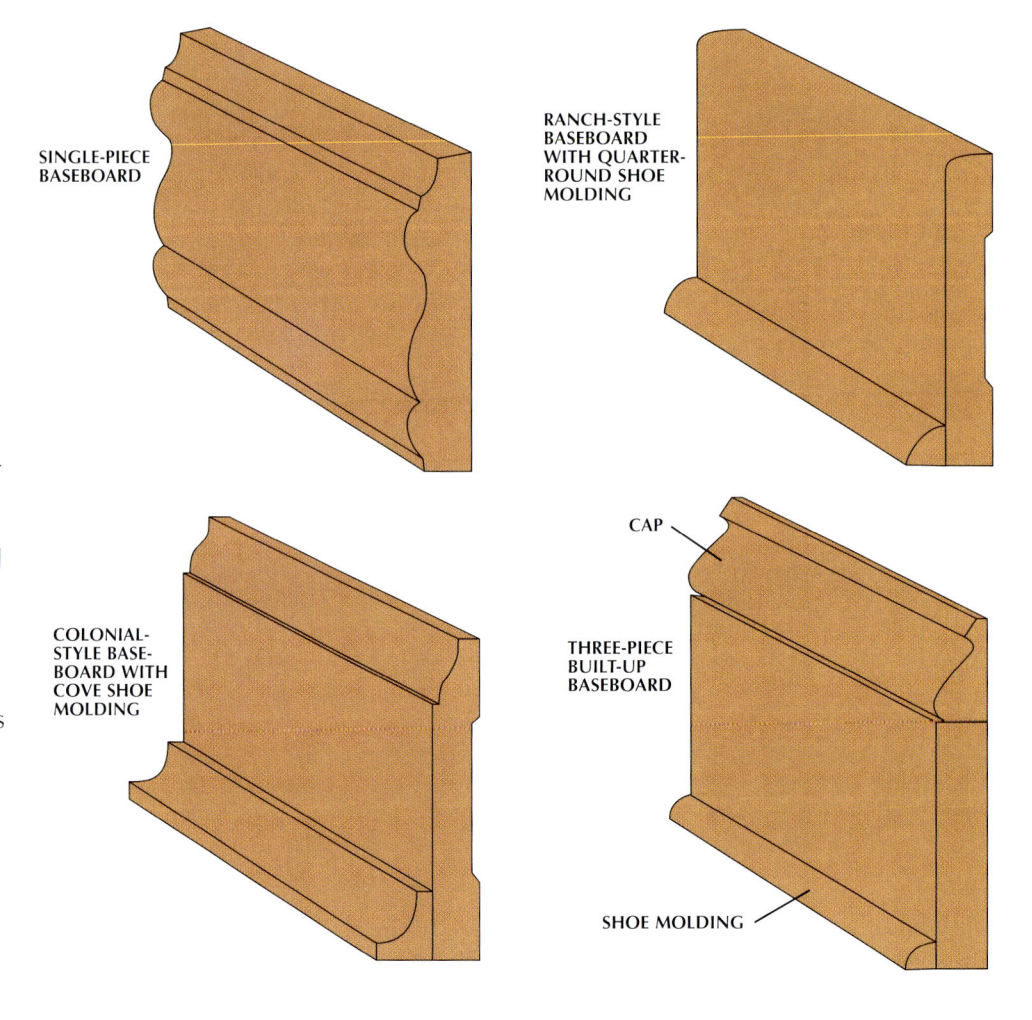

SINGLE-PIECE BASEBOARD

RANCH-STYLE BASEBOARD WITH QUARTER-ROUND SHOE MOLDING

COLONIAL-STYLE BASEBOARD WITH COVE SHOE MOLDING

CAP

THREE-PIECE BUILT-UP BASEBOARD

SHOE MOLDING

PLANNING THE INSTALLATION

Baseboard joinery.
Four different joints may be used when installing baseboard—butt, miter, coped, and scarf *(right)*. Where baseboard meets door casing, it is simply a square butt joint. At outside corners, both pieces have 45-degree miter cuts, or a slightly different angle to fit a wall that is out of square. A butt joint can be used for flat baseboard at an inside corner, but a miter will look neater. For baseboard with a profile, the tightest joint at an inside corner is achieved by cutting one piece square and coping the other to match its contour. To splice baseboard end-to-end along a wall, a scarf joint is used. The pieces are beveled at 45 degrees in the same direction so one wraps over the other.

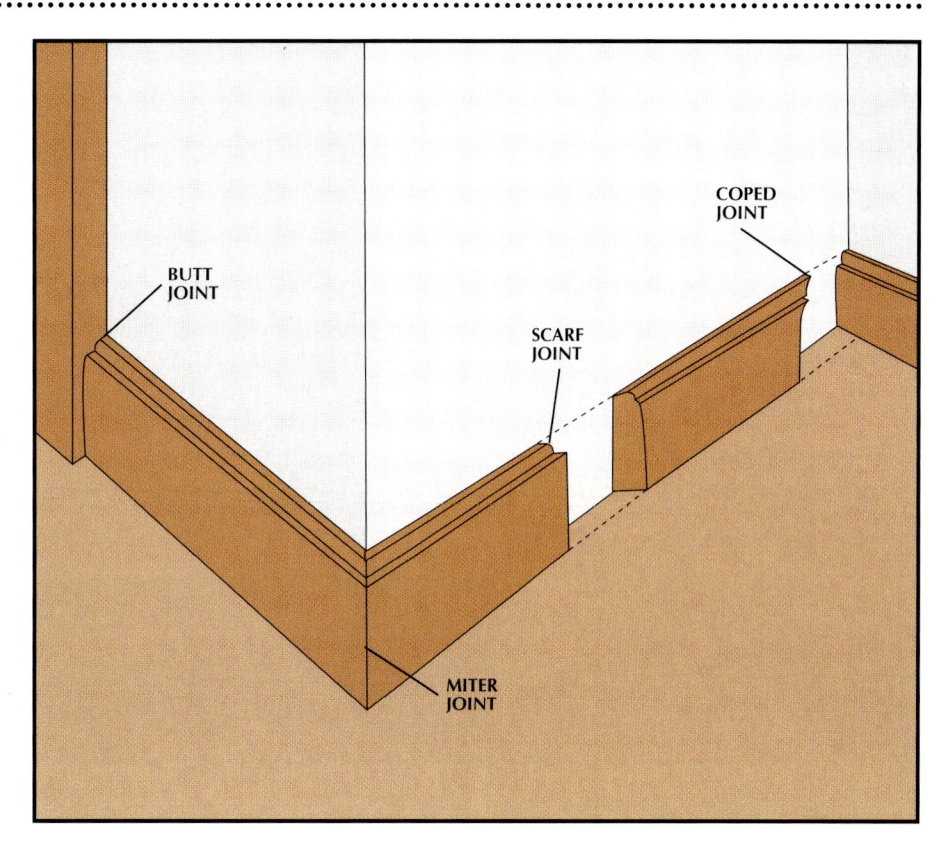

Installation sequence.
Plan the job so coped joints are at right angles to the line of sight from the door. This way the joints will be the least visible if they open up slightly with time. Where possible, make the coped cuts on shorter, easier-to-handle pieces of baseboard.

In the example shown at left, the baseboard on Wall A opposite the door would be installed first, since the trim on the adjoining walls is coped to fit against it. The molding on Wall B would be next for the same reason, then the trim on Wall C to allow the baseboard between it and the door casing to be coped. Angle any scarf joints away from the door so they will be less visible.

MITERING INSIDE CORNERS

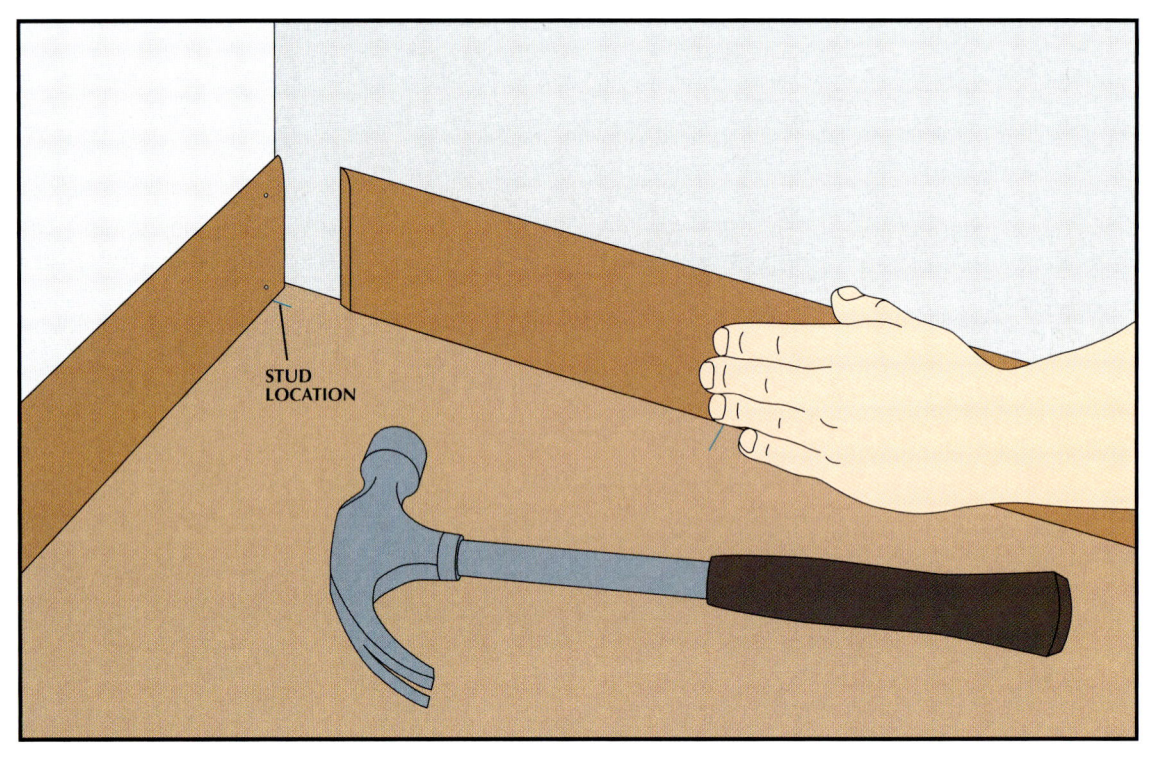

Fitting the corner.
◆ Locate the wall studs with an electronic stud finder and mark their positions on the floor or wall.
◆ With a power or manual miter saw, cut one end of each piece of baseboard at a 45-degree angle.
◆ Test-fit the corner. If there is a gap, shave off a little wood at the back of the joint with a utility knife *(page 51)*.
◆ For softwood stock, attach the first piece with 2-inch finishing nails at each stud location, driving nails $\frac{1}{2}$ inch from the top of the board into the stud and $\frac{1}{2}$ inch from the bottom at an angle into the wall's soleplate. With hardwood baseboards, drill pilot holes for the nails first. If there is no corner stud, fasten the end of the board to the wall with construction adhesive before nailing it to the wall.
◆ Position the mating length of baseboard *(above)* and fasten it in place.
◆ Sink the nails with a nail set.

A FINISH NAILER TO SPEED THE JOB

A power tool called a finish nailer can make the job of installing trim fast and easy. Since the tool drives special thin nails that are less likely to split wood, it is not necessary to drill pilot holes in hardwood. And, unlike a misdirected hammer, a finish nailer will not mar wood. Electric models are available, but pneumatic types like the one featured here are obtainable at tool-rental agencies.

COPING A JOINT

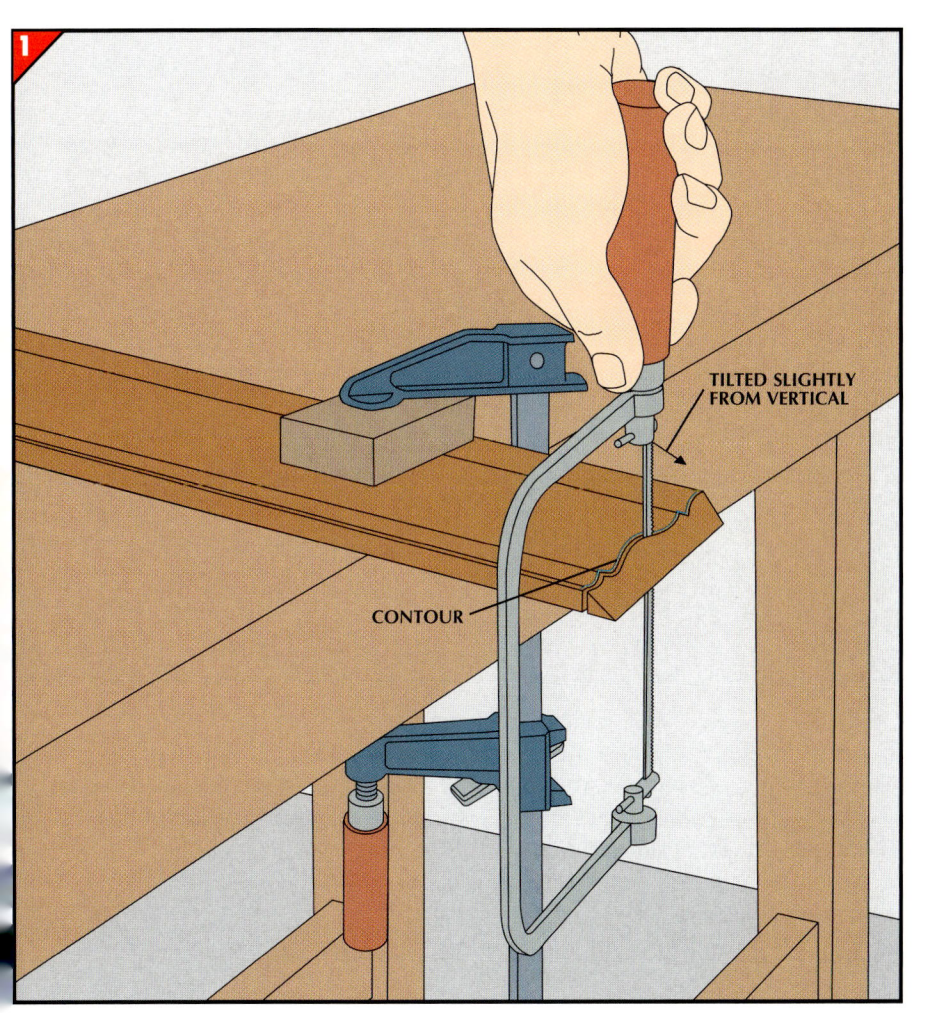

1. Preparing the baseboard.
◆ Cut one piece of molding square so it butts into the corner. Fasten it to the wall as described opposite.
◆ Make a 45-degree miter cut on the end of the second piece and clamp the board face up on a work surface, protecting the stock with a wood pad.
◆ Highlight the exposed contour by tracing it with a pencil.
◆ Fit a coping saw with a narrow blade and cut along the contour, holding the saw upright and slightly tilted to undercut the back of the joint *(left)*. If the blade binds in the kerf, make occasional release cuts into the waste to let small pieces fall away.

2. Installing the coped piece.
◆ Position the coped end of the baseboard against the contour of the first piece to test the fit *(above)*.
◆ Smooth out any irregularities with a round file or fine sandpaper wrapped around a dowel.
◆ Once the fit is perfect, nail the coped molding to the wall studs, then set the nails.

FITTING OUTSIDE CORNERS

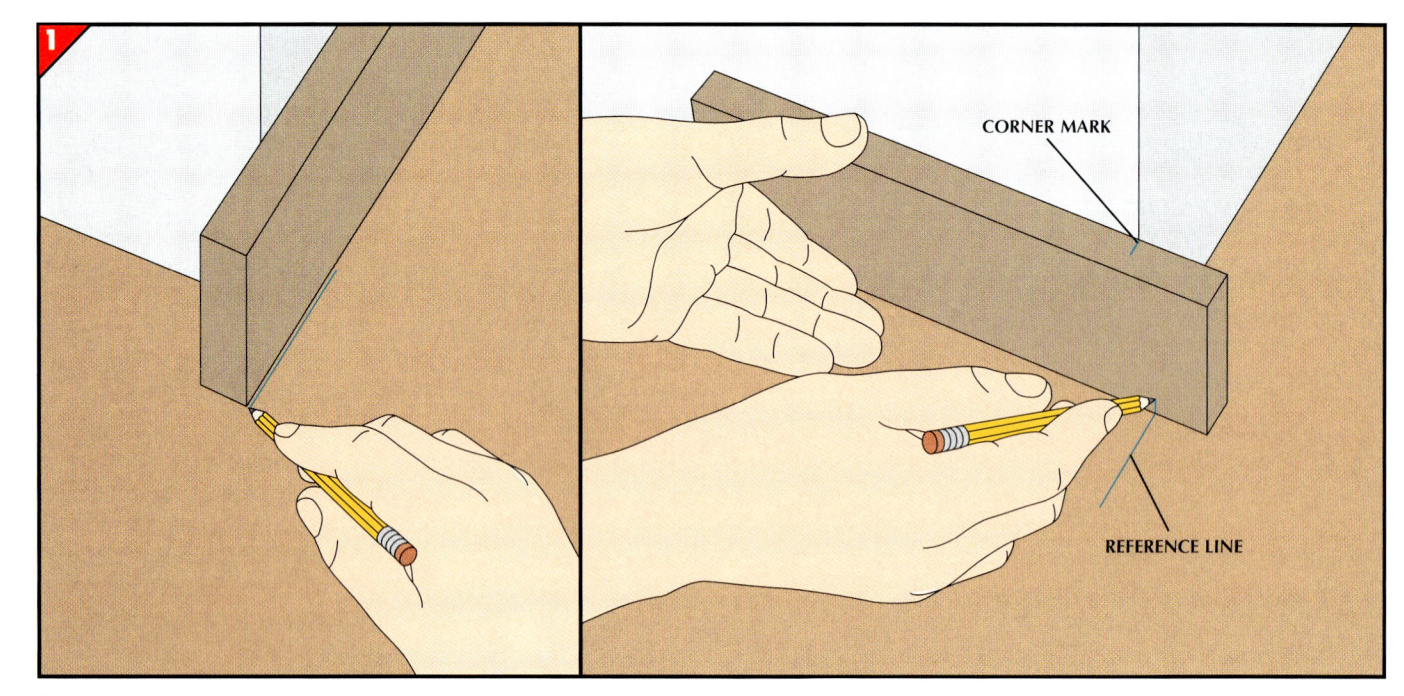

1. Determining the miter angle.
◆ With a long carpenter's square, check the corner of the wall for square. If it is square, proceed to Step 3; otherwise, hold a scrap board the same thickness as the molding against one wall and make a reference mark on the floor along the outside of the board *(above, left)*.
◆ Hold the board against the adjoining wall. Mark the board's top edge in line with the corner and its front face in line with the reference mark on the floor *(above, right)*.

2. Transferring the angle.
◆ With a square, extend the mark on the face of the scrap board to its top edge.
◆ Set an adjustable T-bevel to the angle formed by the end of this line and the corner mark on the top edge of the board *(right)*.

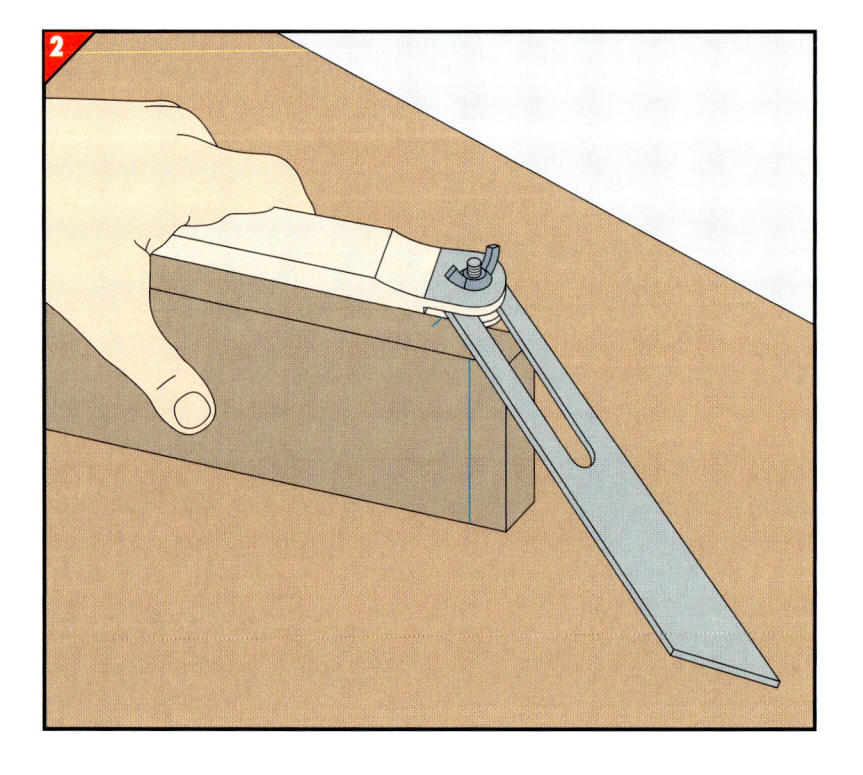

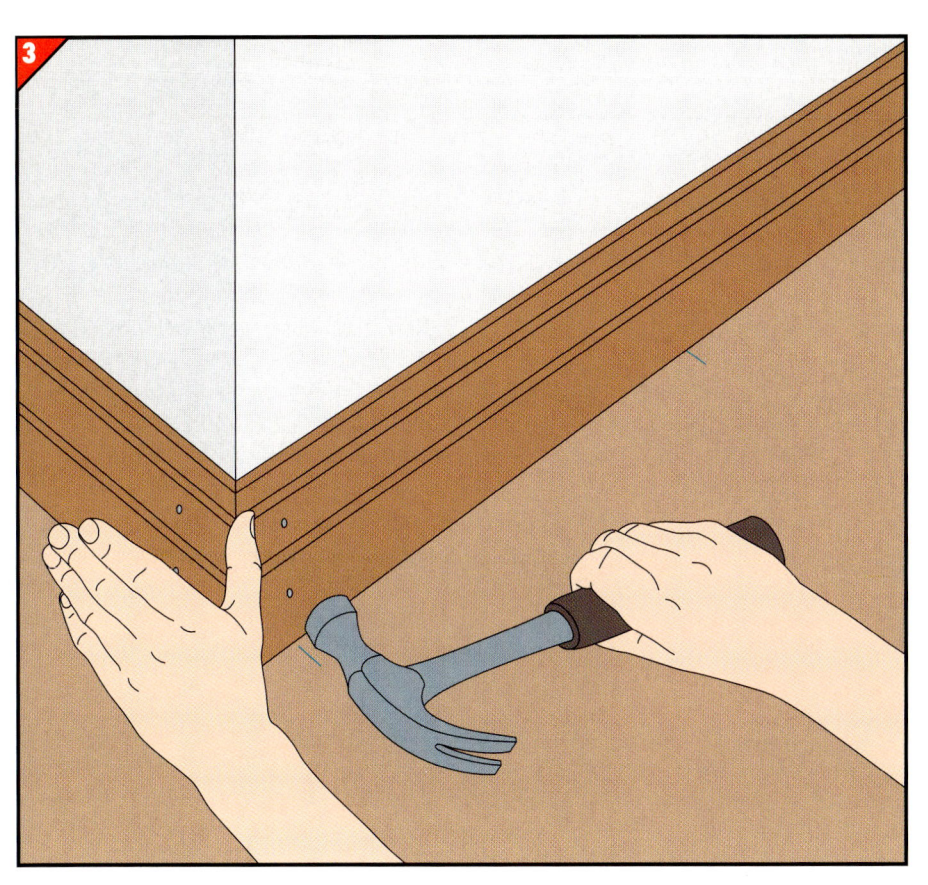

3. Fastening the pieces.
◆ Set the molding on a power or manual miter saw and cut it at 45 degrees. For a wall that is out of square, match the angle measured in Step 2.
◆ Test-fit the joint and make any adjustments by paring away wood with a utility knife *(page 51)* or by sanding.
◆ Fasten the two pieces of molding in the same way as for an inside corner *(page 14)*.

PREFABRICATED CORNER PIECES

Cutting perfect miter joints can be tricky. You can avoid them altogether by purchasing ready-made corner pieces, available from the molding manufacturer for both inside and outside corners *(photographs)*, to match the profile of your baseboard. Once the pieces are installed, adjoining lengths of baseboard can simply be cut square and butted against them.

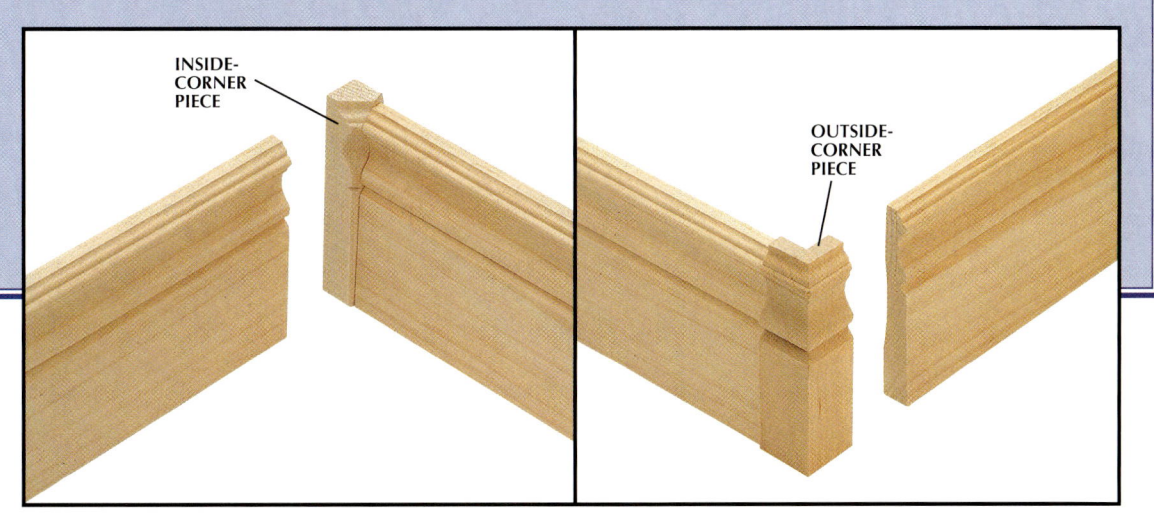

INSIDE-CORNER PIECE

OUTSIDE-CORNER PIECE

SPLICING TRIM FOR A LONG WALL

Nailing a scarf joint.
◆ With a stud finder, locate the wall studs and mark their positions on the floor or wall.
◆ Miter a piece of molding at 45 degrees so the cut will cross a stud, then fasten it in place *(page 14)*, driving two 2-inch finishing nails through the face of the board into the stud and soleplate at the mitered end.
◆ Miter the mating piece so it will mesh with the first piece to continue a straight run. If you will be painting the molding, apply wood glue to the cut ends of both pieces.
◆ Fit the second piece in place *(left)* and fasten it so the nails at the cut end just clear the joint *(inset)*.
◆ Sink all the nails.

TRICKS OF THE TRADE

Fitting Baseboard at Doors

A shop-made jig called a preacher will help you fit baseboard tightly against door casing. Make the jig by cutting a 1-by-6 about 1 foot long, then notch one edge as high as the baseboard and as wide as the casing and baseboard combined. To run baseboard to the door casing, first cut the opposite end to fit the wall or corner, then prop the piece in place. Hold the preacher over the baseboard with the notched edge flush against the outside edge of the casing, then mark the baseboard along the notch *(right)*.

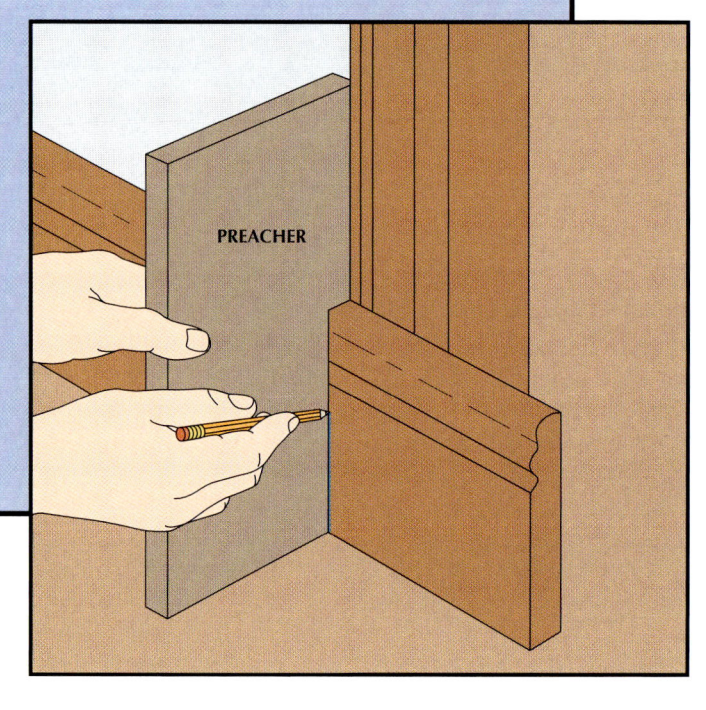

ADDING SHOE MOLDING

Nailing the pieces.
- Fit shoe molding at corners in the same way as for baseboard: with miters or coped joints at inside corners *(pages 14-15)* and miters at outside corners *(page 16)*.
- Fasten the molding to the floor with $1\frac{1}{2}$-inch nails driven every 16 inches *(left)*.
- Sink the nails with a nail set.

CAPPING OFF THE JOB

For a more decorative effect and to hide irregularities in the wall surface, you can install baseboard with a cap piece. Some caps are made to mesh to baseboard in a tongue-and-groove joint. Others, as shown at right, are fastened with $1\frac{1}{2}$-inch finishing nails driven through the cap into the baseboard every 16 inches. The cap is fitted at corners in the same way as baseboard—with butt, miter, and coped joints. Shoe molding completes the installation.

BASEBOARD CAP
BASEBOARD
SHOE MOLDING

Installing Crown Molding

Originally designed to hide joints between the tops of walls and the ceiling, crown molding lends a formal look to a room. Typically installed in one-piece lengths, it also can be combined with other trim for a more ornate effect *(page 23)*.

Molding Size: Buy stock that is properly proportioned for the room; molding that is too wide will make the ceiling appear lower than it is. For a standard 8-foot-high ceiling, for instance, $3\frac{1}{2}$-inch crown is appropriate. To aid in installation, mark a chalk line on the wall to indicate where the molding will sit; because not all crown sits at the same angle, you must measure the "projection" of the molding before you can make the chalk line *(below)*.

Cutting the Pieces: Crown molding is fitted in much the same way as baseboard—outside corners are mitered *(opposite)*, inside corners are mitered or coped *(pages 22-23)*, and lengths of molding are joined end-to-end with scarf joints *(page 18)*. But since crown is installed at an angle, all of these cuts are compound cuts. To make the cuts accurately, set the molding in the saw upside down, as if the base of the saw were the ceiling and the fence were the wall.

Fastening the Trim: Where the ceiling joists run at right angles to the wall, crown molding is nailed through its flat portions along the edges into the wall studs and joists. If the joists are parallel to the wall, nailing the molding to the studs and gluing it to the ceiling is sufficient. A nail driven into the ceiling at each end and the middle of the molding will hold the trim in place until the glue dries.

Sink all the nails with a nail set, then cover the holes with spackling compound if you intend to paint the moldings, or with wood putty if a clear finish will be applied.

TOOLS
Carpenter's square
Chalk line
Electronic stud finder
Adjustable T-bevel
Hammer
Power or manual miter saw
Electric drill
Vise
Coping saw
Round file
Nail set

MATERIALS
Crown molding
Finishing nails ($1\frac{1}{4}$", $2\frac{1}{2}$")
Wood glue
Spackling compound or wood putty

SAFETY TIPS
Protect your eyes with goggles when hammering or when using a power saw.

MITERING AT OUTSIDE CORNERS

1. Measuring the projection.
◆ Cut a small piece of molding and hold it against a carpenter's square, oriented as it will be in the room, so the flat sections of the trim are flat against the arms of the square *(right)*. Note the distance that the molding projects horizontally (equivalent to the distance along the ceiling) and vertically (the distance along the wall).
◆ At each corner of the room, measure from the ceiling a distance equal to the vertical projection and make a mark on the wall. Snap a chalk line all the way around the room at these points.
◆ With an electronic stud finder, locate and mark the studs and ceiling joists.

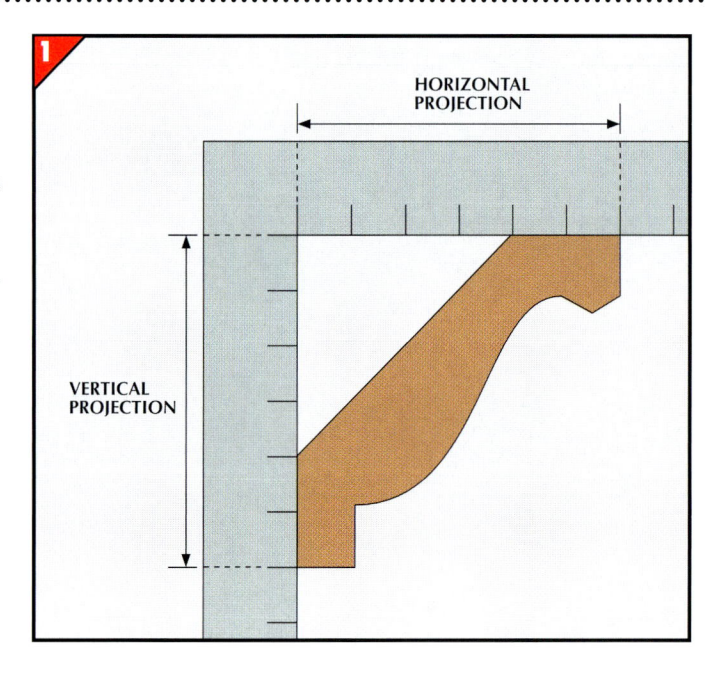

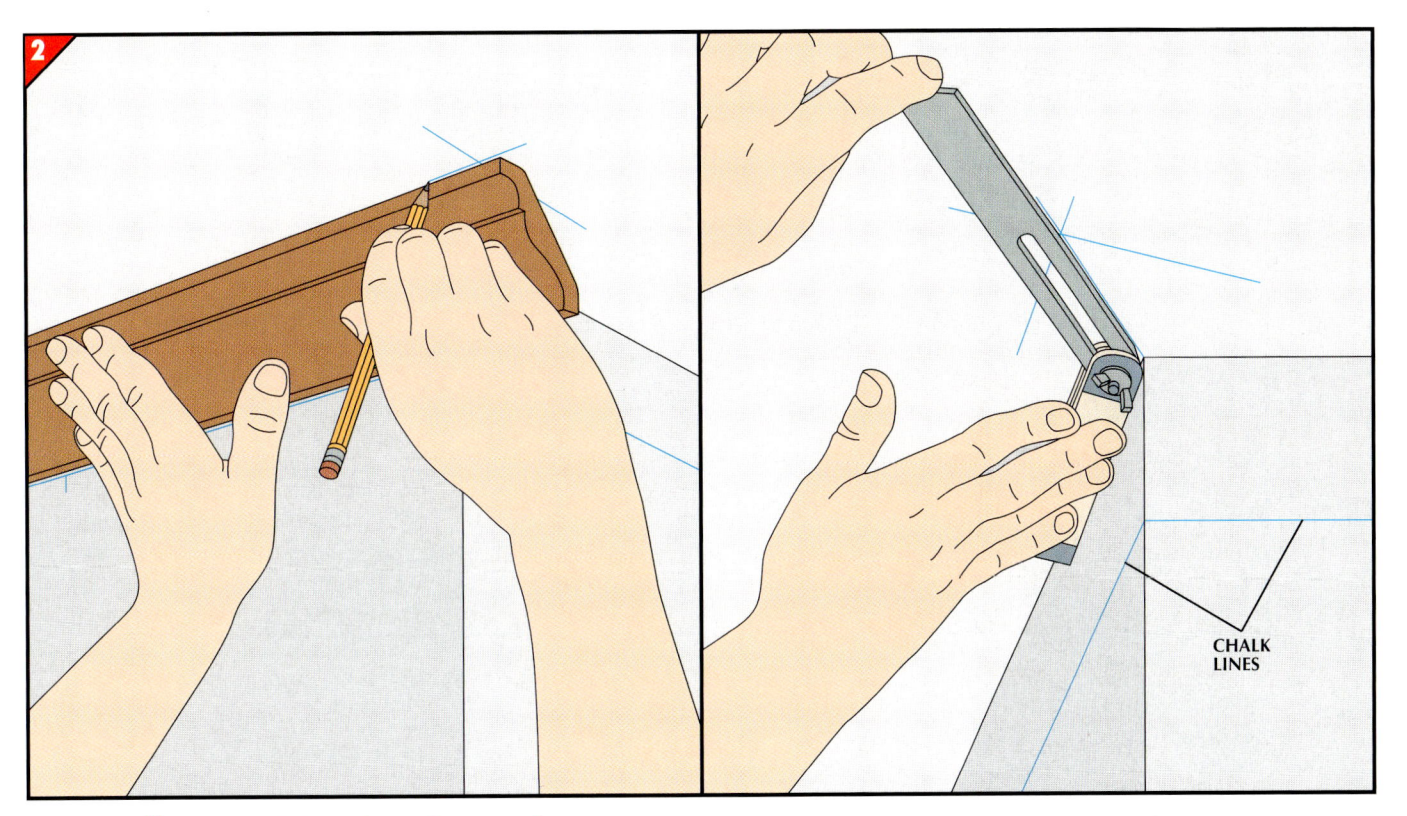

2. Determining the miter angle.

◆ Position a piece of molding against one wall, aligning the bottom edge with the chalk line; make a mark at the top corner, extending it past the corner. Repeat on an adjacent wall to make a second mark that crosses the first *(above, left)*.

◆ Draw a third line from the corner to the intersection of the first two lines.
◆ Adjust a T-bevel so its handle butts against one of the walls and the blade aligns with the third line *(above, right)*.

3. Making a compound cut.

◆ Set the blade of a power or manual miter saw to the angle measured in Step 2. If the corner is perfectly square, this angle will be 45 degrees.
◆ Stick a strip of masking tape to the saw base, measure out from the saw fence a distance equal to the horizontal projection of the crown molding *(Step 1)*, and mark a line on the tape parallel to the fence.
◆ Set the molding on the saw upside down so its top edge—which will rest against the ceiling—is flush against the base. Place molding that will be installed to the right of the corner to the left of the saw blade and vice versa. Support a long piece of molding with blocks of wood.
◆ Holding the molding firmly against the fence with its edge even with the line on the masking tape, keep your hand well away from the blade and make the cut *(left)*.

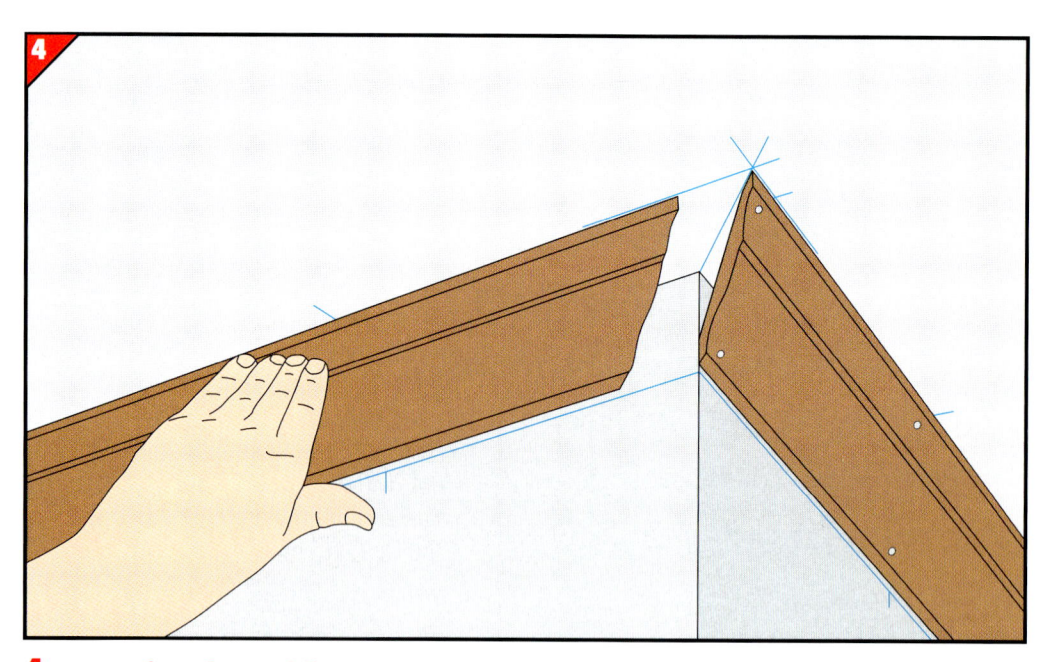

4. Fastening the molding.

◆ Position one piece of molding, aligning the bottom edge with the chalk line and the mitered corner with the point where the three marked lines *(Step 2)* cross. Tack it in place.
◆ Set the second piece of crown in place and test the fit. If necessary, take both pieces down and pare the cuts with a utility knife *(page 51)*.
◆ Fasten one piece of molding with $2\frac{1}{2}$-inch finishing nails into the wall studs and ceiling joists, first drilling pilot holes if the molding is hardwood.
◆ Apply wood glue to the mating ends of both pieces of molding and position the second piece *(above)*. Fasten it in the same way as the first.
◆ Drill a pilot hole for a $1\frac{1}{4}$-inch finishing nail through the joint from each side and drive the nails.

COPING INSIDE CORNERS

1. Making the cope cut.

◆ Measure the projection of the crown molding and snap chalk lines on the wall *(page 20, Step 1)*.
◆ Miter one end of a piece of crown molding at 45 degrees.
◆ Clamp the molding in a vise in the same orientation it will be when installed, then run a pencil along the exposed contour to highlight it.
◆ Cut along the contour with a coping saw, holding the saw almost perpendicular to the molding but angled very slightly toward the back of the piece to undercut the joint *(right)*.

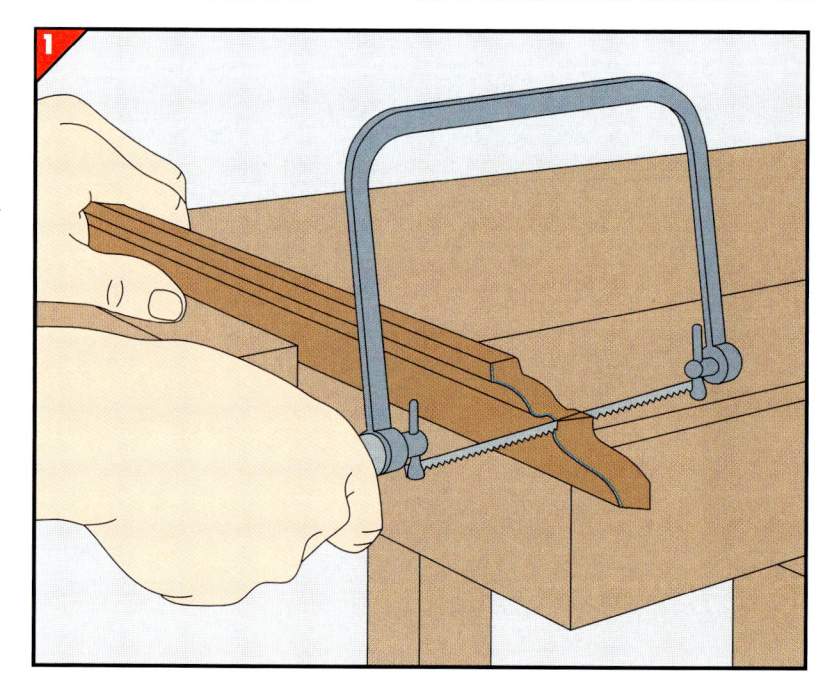

2. Installing the molding.

◆ Cut a length of crown molding square to fit at the corner, then fasten it in place *(opposite, Step 4)*.
◆ Test-fit the coped piece against the molding *(left)*. If necessary, shape the coped piece with a round file or some fine sandpaper wrapped around a dowel.
◆ Fasten the coped piece in place.
◆ Sink all the nails with a nail set.

ELABORATE CROWN STYLES

You can achieve an ornate look by combining single-piece crown molding with two pieces of flat molding such as baseboard *(right, top)*. First, install the flat moldings on the wall and ceiling with a slight gap between them to allow for expansion of the wood. Then, fasten standard crown molding to the flat molding as you would to the wall and ceiling.

A traditional formal cornice *(right, bottom)* is best suited to a room with high ceilings. To install this trim, first fasten nailer strips to the wall and ceiling with a gap between them. Complete the box with a soffit and fascia cut from $\frac{3}{4}$-inch stock, shaping a lip along the bottom of the fascia on a router table *(page 11)*, then add wall molding directly below it. Install crown molding between the fascia and the ceiling, then add a smaller crown—known as bed molding—between the soffit and the wall molding.

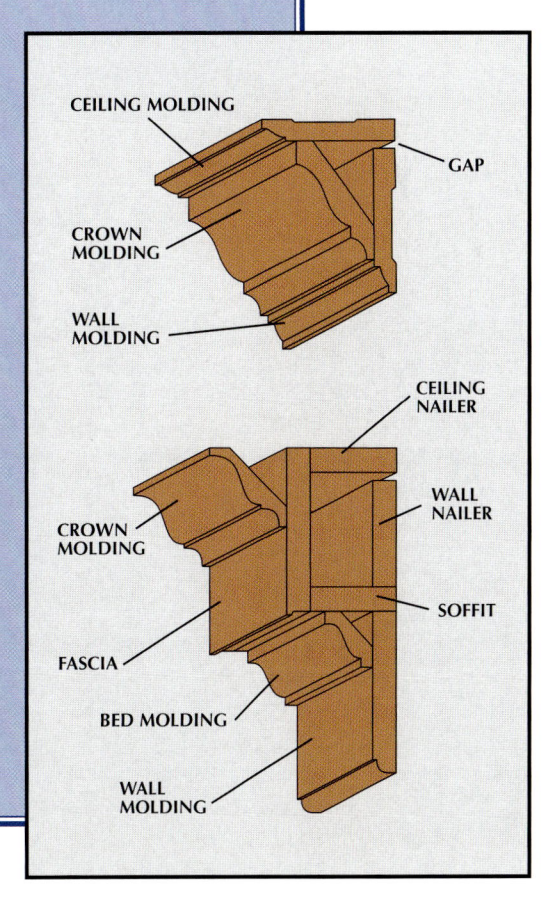

Embellishing Walls with Rails

Chair and picture rail are traditional styles of trim installed partway up the wall. Available in a variety of profiles *(below)*, they are fastened in the same way as baseboard, with miters at outside corners; miters or coped joints at inside ones; and scarf joints at splices on long walls. Although chair and picture rail are seldom used together, either can be combined with baseboard and crown moldings.

Chair Rail: Originally designed to prevent chair backs from marring walls or paneling, chair rails are now primarily decorative, often used to divide types of wall treatments. The area above the rail is usually painted, with paper or paneling covering the portion below. Install chair rail about one-third of the way up the wall, typically at a height of 3 feet, positioned parallel to the floor, whether or not the floor is level *(opposite)*.

Picture Rail: Like chair rail, this molding is now used mostly for decoration. If you plan to hang frames from the trim, however, angle the nails down slightly when fastening it the wall to strengthen the connection and help the molding support the weight. Install picture rail in the same way as chair rail, but position it about 2 feet below the ceiling along a chalk line that follows the ceiling.

TOOLS

Chalk line
Electronic stud finder
Power or manual miter saw
Electric drill
Coping saw
Hammer
Nail set

MATERIALS

Chair rail
Picture rail
Finishing nails (2")
Spackling compound or wood putty

SAFETY TIPS

Goggles protect your eyes when you are hammering or are operating a power tool.

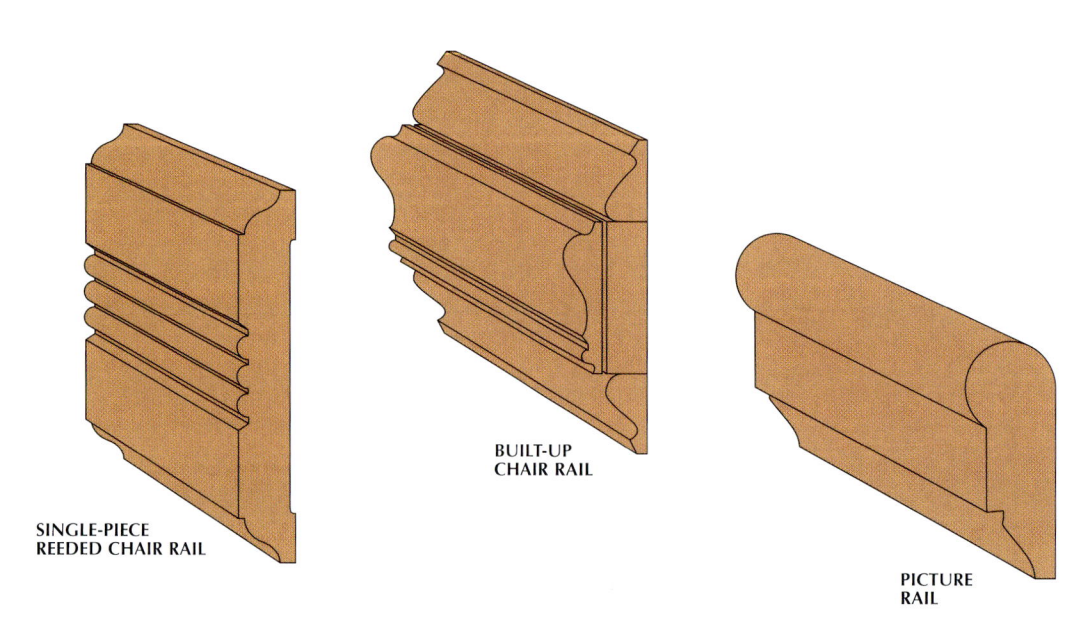

SINGLE-PIECE REEDED CHAIR RAIL

BUILT-UP CHAIR RAIL

PICTURE RAIL

Rail styles.

Simple one-piece chair rail comes in a variety of styles, often with a reeded profile. For a more complex profile, the trim can also be built up of more than one piece. The rounded lip along the top edge of picture rail was originally designed to support the hooks and wires that hold up picture frames.

FASTENING CHAIR RAIL

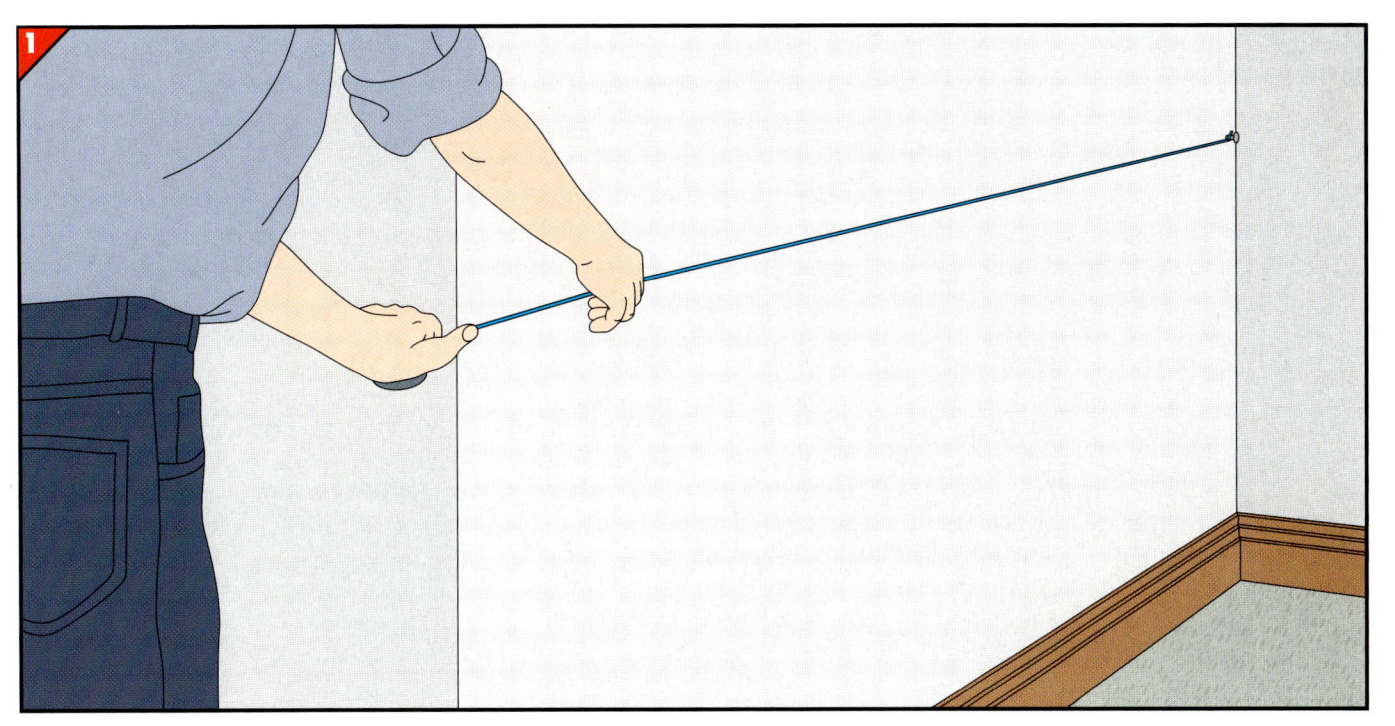

1. Snapping a guideline.
- Make a mark at each end of one wall 36 inches above the floor.
- Drive a finishing nail into the wall at one mark and hook a chalk line on the nailhead. Align the opposite end of the chalk line with the second mark and snap the line *(above)*.
- Mark the location of the chair rail on the other walls in the same way.
- With an electronic stud finder, locate and mark the wall studs.

2. Nailing the rail to the wall.
- Cut the chair rail to length, mitering or coping it at corners as you would baseboard *(pages 12-18)*.
- Align the top edge of the trim with the chalk line.
- For hardwood, drill pilot holes for two 2-inch finishing nails at each stud mark, then drive the nails through the rail into the studs *(left)*.
- Sink the nailheads with a nail set. If you will be painting the rail, fill the holes with spackling compound; otherwise, use wood putty.

Tongue-and-Groove Wainscoting

Paneling installed on the bottom portion of a wall, referred to as wainscoting, adds warmth and character to a room. It can also make a space appear smaller and busier, so it should be reserved for rooms with high ceilings and few windows.

Choosing Lumber: Simple to install, tongue-and-groove wainscoting comes in either paint-grade softwood or stain-grade hardwood. In addition to flat pieces, you can choose from a variety of beaded and beveled styles. Available in thicknesses of $\frac{5}{16}$ and $\frac{3}{8}$ inch, boards for wainscoting are most convenient precut to length—36 inches—to avoid having to cut panels from longer pieces.

Planning the Installation: Start by measuring the wall to determine whether you will need to trim a board to width to fit at one end, and install the slightly narrower piece at the less noticeable end of the wall. But rather than cutting an end panel very thin, trim the boards at each end of the wall by the same amount.

Where a room includes an outside corner, begin there—miter the corner boards or set them flush against each other and cover the joint with corner molding. At inside corners, butt the pieces against each other.

If the floor is uneven, cut the boards a little short so you can install them with their tops at the same level; gaps at the bottom will be concealed by baseboard. Since gaps may open up between the panels over time, stain or paint the boards before you install them.

Tongue-and-groove boards can be glued to wallboard; for plaster walls, first install furring strips *(opposite)*.

TOOLS
Chalk line
Carpenter's square
Table saw
Hammer
Electric drill
Nail set
Carpenter's level
Jack plane
Router

MATERIALS
Tongue-and-groove wainscoting
Lumber ($\frac{3}{4}$")
Finishing nails ($1\frac{1}{2}$", 2")
Panel adhesive
Spackling compound or wood putty

SAFETY TIPS
When hammering or when using a power tool, protect your eyes with goggles.

A paneled wall.
In this typical tongue-and-groove wainscoting, the groove of each successive piece fits over the tongue of the previous one so the shoulder along the back presses snugly against the grooved edge. A bead at the front adds an attractive profile. Baseboard covers gaps at the bottom of the boards and a bullnose cap hides their top edges.

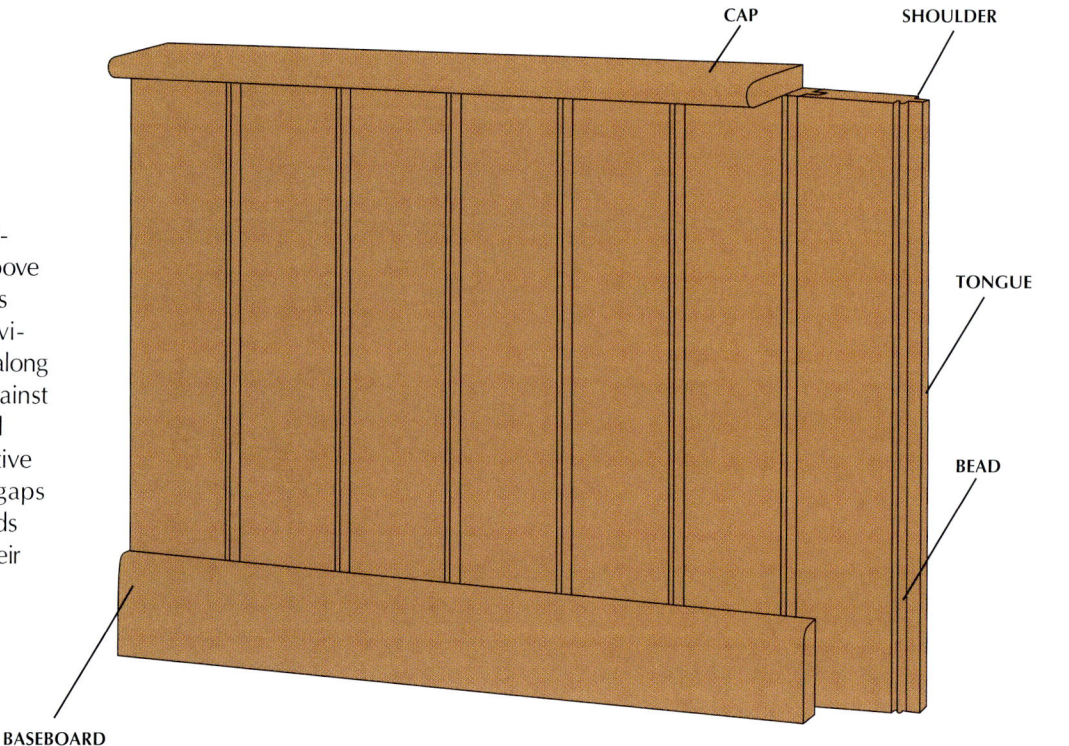

Labels: CAP, SHOULDER, TONGUE, BEAD, BASEBOARD

FURRING STRIPS FOR PLASTER WALLS

Because tongue-and-groove boards will not adhere well when glued to plaster walls, you'll need to nail four rows of 1-by-3 furring strips to the studs first, then fasten the paneling to them. Mark stud locations along the installation route, then attach one strip even with the top of the boards and another one along the floor, driving two common nails at each stud long enough to penetrate the studs by $1\frac{1}{2}$ inches. Add two more strips evenly spaced between the first two. Install the boards as on wallboard *(below)*, nailing them to the furring strips.

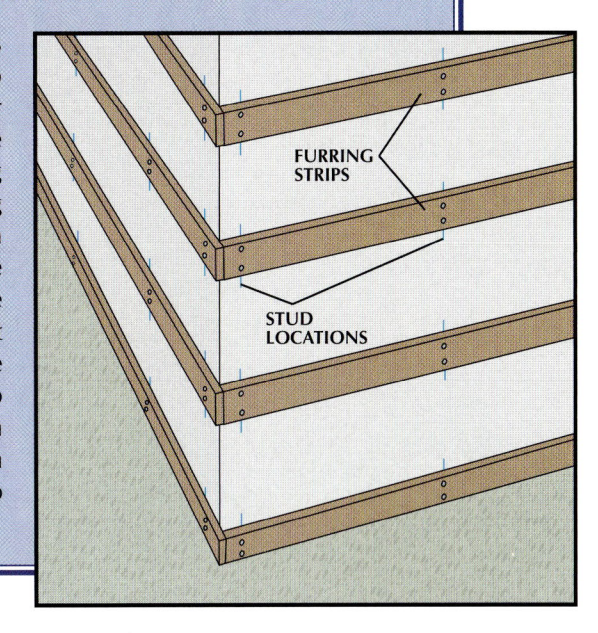

FITTING AND NAILING THE BOARDS

1. Starting at an outside corner.
◆ Snap a chalk line parallel to the floor at a height of 36 inches *(page 25, Step 1)*.
◆ Determine the miter angle to be cut at the corner as for baseboard *(page 16, Step 1)*, and on a table saw, miter the grooved edge of each corner board.
◆ Apply panel adhesive to the back of one board, position it at the corner so its top edge aligns with the chalk line, and press it in place.
◆ To secure the board while the adhesive dries, drive two $1\frac{1}{2}$-inch finishing nails near the top and bottom through the tongue, angling them so they pass through the back of the board rather than the back of the tongue; drill pilot holes first if the boards are hardwood.
◆ Fasten the adjacent corner board in the same way, holding the miters together *(left)* and driving four nails through the face of each board along the mitered edges.
◆ With a nail set, sink the nails driven through the tongues flush with the surface. Set those along the mitered edge below the surface.

2. Paneling the wall.

◆ Apply adhesive to the back of the next board and slip its grooved edge over the tongue of one of the corner boards, aligning the top edge with the chalk line.
◆ Nail the board in place (Step 1).
◆ Fasten four or five boards along the wall in the same way (left).

TRICKS OF THE TRADE

A Block for a Snug Fit

Tongue-and-groove boards sometimes fit so tightly that a hammer blow is needed to close the joint. To avoid damaging the tongue of a board with the hammer, cut a short length of wainscoting as a hammering block, slip it over the tongue of the stubborn piece, and strike the block to force the board into place (right).

3. Checking for plumb.
◆ Hold a carpenter's level against the tongue of the last board installed to check for plumb *(left)*. If the board is not perfectly vertical, taper the grooved edge of the next few boards slightly with a jack plane to bring the paneling back to the vertical.
◆ Install the remaining pieces, adjusting for plumb every few boards, until there is room for only one more board.

4. Turning a corner.
◆ To install the last board along the wall —at an inside corner or a door opening —trim the tongued edge to fit, beveling it slightly. Fit the groove over the tongue of the last board installed, then swing the beveled edge into place. Fasten this board to the wall with four nails driven along the beveled edge.
◆ Begin paneling the adjoining wall by butting the grooved edge of a board against the face of the last board installed *(right)*. Fasten it to the wall along the tongue and the grooved edge.
◆ Continue fastening the rest of the paneling around the remaining walls.

5. Adding a cap.
◆ To make the cap, round one edge of a piece of ¾-inch lumber with a router *(page 11)*, then, on a table saw, rip the piece about 1 inch wide.
◆ Position the cap on top of the wainscoting *(right)* and toenail it to the wall studs with 2-inch finishing nails.
◆ Continue to fasten cap stock to the paneling, joining lengths end-to-end with scarf joints *(page 18)* and mitering pieces at inside and outside corners.
◆ Install baseboard *(pages 12-18)*.
◆ Sink the nails with a nail set. If you plan to paint the paneling, fill the holes with spackling compound; otherwise use wood putty.

DECORATIVE TOUCHES

To dress up wainscoting, you can make the cap stock wider than described above and add a strip of chair-rail molding directly below the cap piece *(right, top)*. Alternatively, buy prefabricated wainscoting cap *(right, bottom)*, which covers the tops of the boards and adds an attractive profile on the front of the paneling.

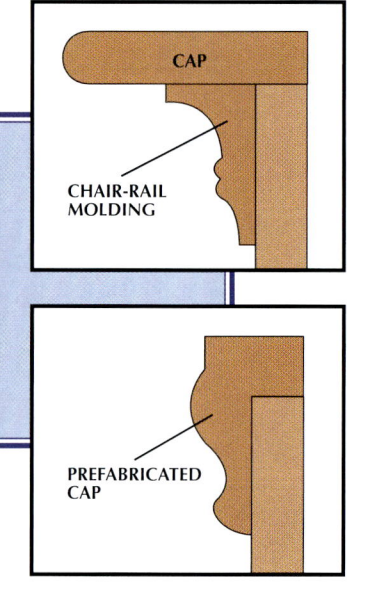

Frame-and-Panel Wainscoting

The most traditional style of wainscoting is frame-and-panel. Although more time-consuming to install than tongue-and-groove, it lends a formal elegance to a room. Originally constructed in the same painstaking way as a frame-and-panel door *(pages 60-64)*, the wainscoting can be built more simply by attaching stiles, rails, and panels to a plywood backing.

Planning: Begin the job by making a scale drawing of the walls, showing the locations of door and window openings and electrical outlets. Then, superimpose a sketch of the paneling, following the guidelines below and basing the size of the elements on the proportions of the wall. If possible, plan the layout so holes for electrical boxes fall on a flat surface rather than on a joint or a section of molding; otherwise, you'll have to move the boxes.

Materials and Finishing: The base and all of the decorative elements can be made of furniture-grade plywood. If you plan to stain the wainscoting or apply a clear finish, choose a type with an attractive hardwood veneer.

TOOLS
Chalk line
Electronic stud finder
Carpenter's level
Circular saw
Electric drill
Hammer
Nail set
Table saw
Power or manual miter saw

MATERIALS
Furniture-grade plywood ($\frac{1}{4}$", $\frac{1}{2}$")
Lumber (1")
Picture-frame molding
Baseboard
Panel adhesive
Finishing nails (1", 2", $2\frac{1}{2}$")
Paneling nails ($1\frac{5}{8}$")
Mineral spirits
Wood putty or spackling compound

SAFETY TIPS
Wear goggles when operating a power tool or driving nails.

Imitating frames and panels.

Frame-and-panel wainscoting usually extends about 36 inches from floor to cap. It can be made from furniture-grade plywood, with $\frac{1}{2}$-inch-thick stiles and rails glued to a $\frac{1}{4}$-inch-thick base sitting $\frac{1}{4}$ inch off the floor. Stiles range from 4 to 6 inches wide and are positioned to cover the joints between the base panels. The top rail is generally $2\frac{1}{2}$ to $3\frac{1}{2}$ inches wide, with the visible part of the bottom rail being the same or up to 2 inches wider. In addition, 1 inch of the bottom rail supports the baseboard, which also has a backing strip behind it along the floor. The rectangle formed by the stiles and rails should not exceed 24 inches—except below windows where it is wide enough to span the opening—and has a 2-inch gap between it and the surrounding rails and stiles. All the exposed edges of plywood are concealed with picture-frame trim or quarter-round molding, and the top rail is covered with a cap.

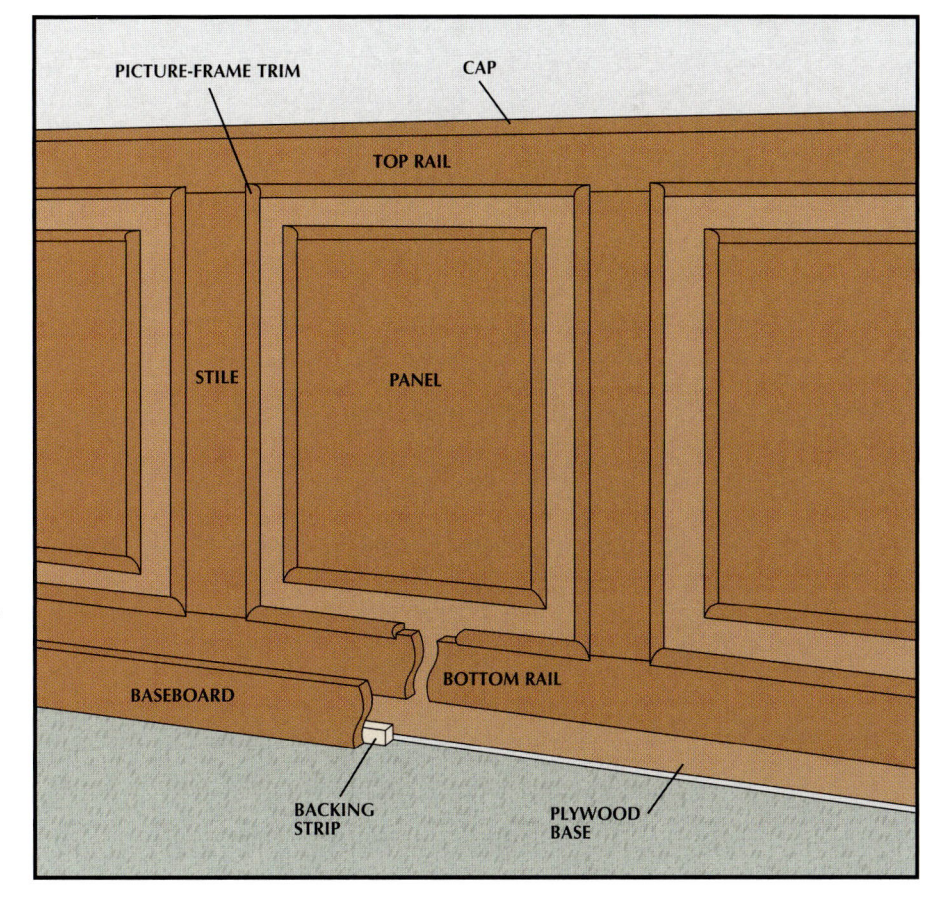

ATTACHING THE PLYWOOD BASE

1. Applying adhesive.
◆ With a circular or table saw, cut a $\frac{1}{4}$-inch furniture-grade plywood panel to a height of 36 inches.
◆ Snap a chalk line parallel to the floor at a height of $36\frac{1}{4}$ inches *(page 25)*.
◆ Locate and mark the wall studs with an electronic stud finder. Use a carpenter's level to mark a plumb line at each stud location, extending it a few inches above the chalk line.
◆ Apply a $\frac{1}{8}$-inch-wide bead of panel adhesive in an S-pattern just below the chalk line, above the floor, and along the plumb lines *(right)*.

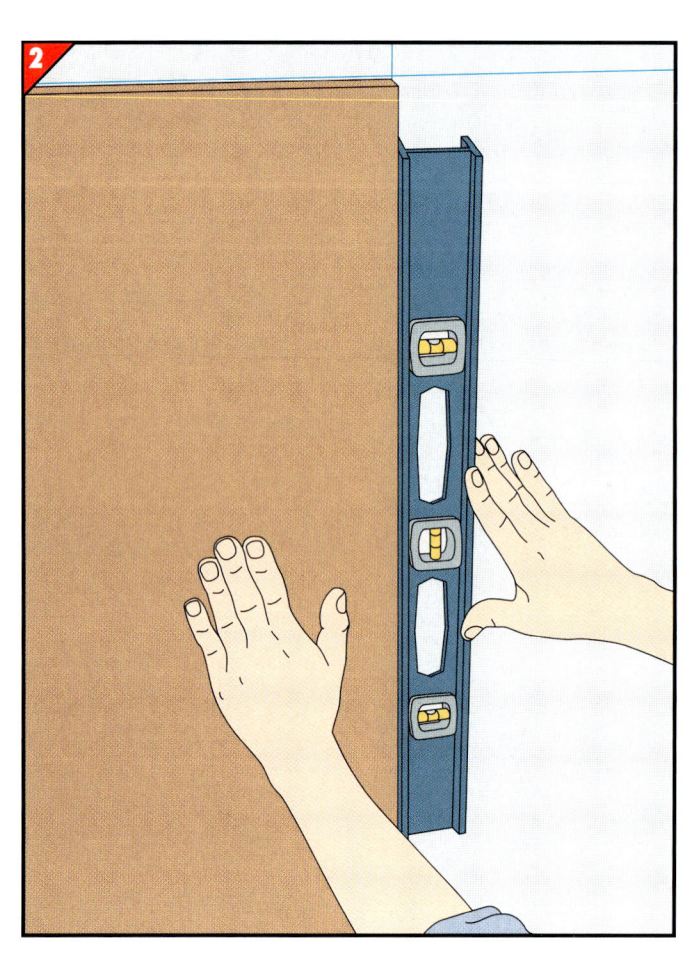

2. Plumbing the first panel.
◆ Place $\frac{1}{4}$-inch spacers on the floor against the wall.
◆ Resting the panel on the spacers, press it lightly against the adhesive.
◆ With a carpenter's level, check that the panel is plumb *(left)*. Add shims, if necessary, to level the panel.
◆ Secure the top edge of the panel to the studs with $1\frac{5}{8}$-inch paneling nails—you may need longer nails if you are fastening through plaster and lath.

3. Securing the panel.

◆ Press the panel against the wall to compress the adhesive; if specified by the adhesive directions, pull out the bottom of the panel and insert a 6-inch wood block at each end between the panel and the wall *(right)*. Let the adhesive dry according to the manufacturer's instructions, then remove the blocks and push the panel against the wall again, tapping it with your fist to make a tight seal.
◆ Nail the bottom edge of the panel to the wall at the studs; fasten the ends to the studs with a vertical row of nails driven at 1-foot intervals.
◆ With a cloth dipped in mineral spirits, wipe away any adhesive on the face of the panel.
◆ Install the remaining panels end-to-end along the wall in the same way.
◆ Sink all the nails with a nail set.

⚠ **CAUTION** *To prevent spontaneous combustion of cloths soaked in mineral spirits, hang them outside to dry, or store them in an airtight metal or glass container.*

BUILDING UP THE SURFACE

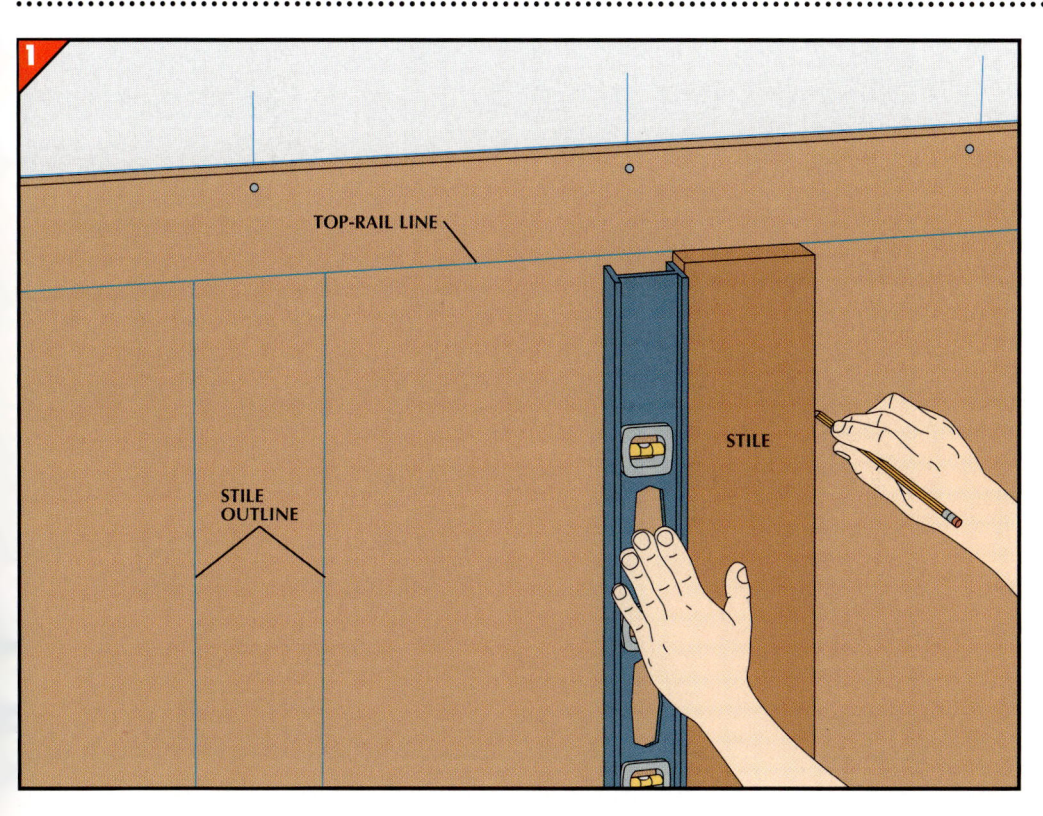

1. Marking layout lines.

◆ Draw a level line near the top of the panels, outlining the position of the bottom of the top rail.
◆ Mark a level line for the bottom edge of the bottom rail 1 inch lower than the height of the baseboard to be installed.
◆ Cut a $\frac{1}{2}$-inch plywood stile to width, long enough to span from the top-rail line to the bottom-rail line.
◆ Holding the stile plumb in position, mark a line on each side of it, then repeat the procedure to mark all the stile locations *(left)*.

2. Attaching the stiles and rails.

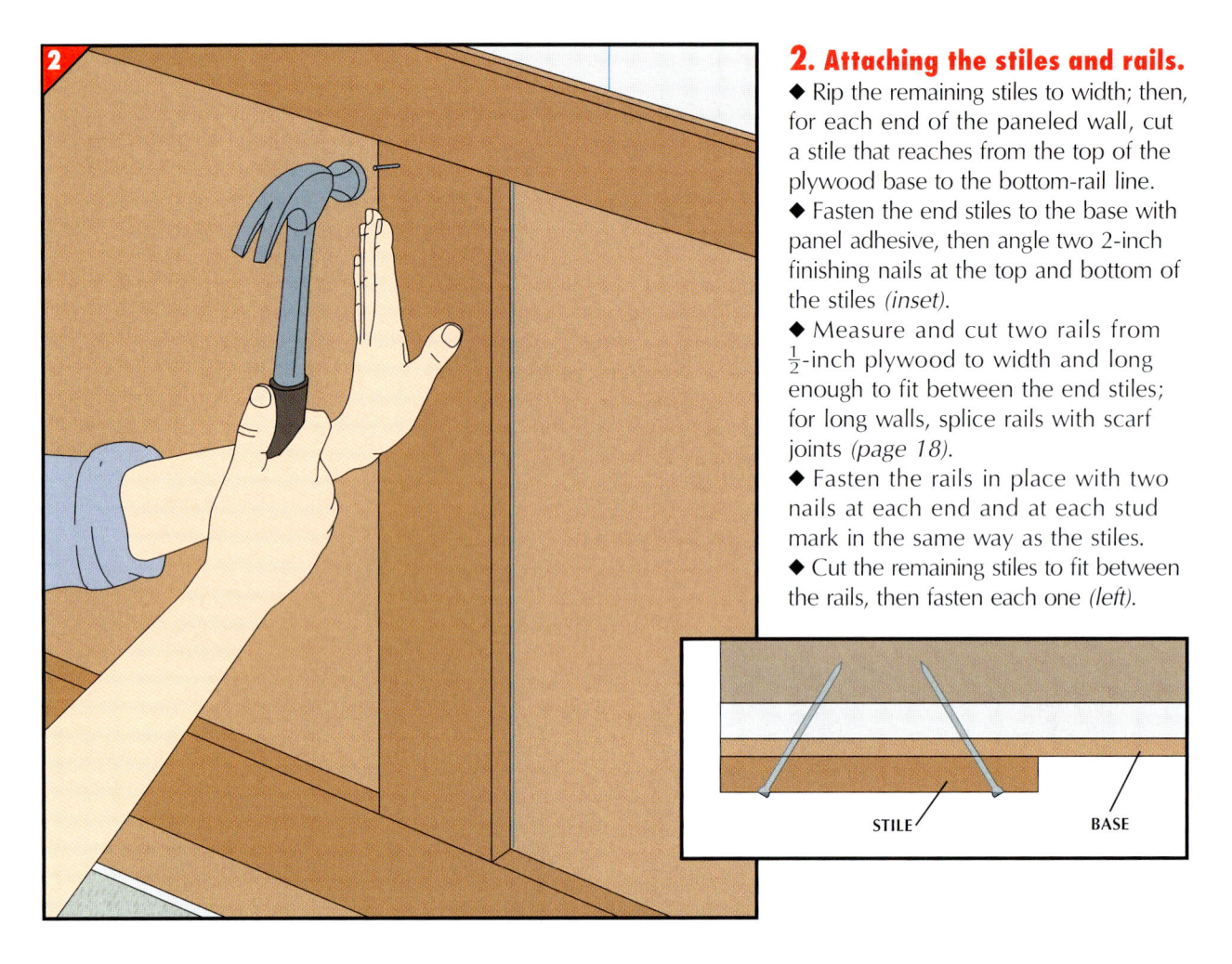

◆ Rip the remaining stiles to width; then, for each end of the paneled wall, cut a stile that reaches from the top of the plywood base to the bottom-rail line.
◆ Fasten the end stiles to the base with panel adhesive, then angle two 2-inch finishing nails at the top and bottom of the stiles *(inset)*.
◆ Measure and cut two rails from $\frac{1}{2}$-inch plywood to width and long enough to fit between the end stiles; for long walls, splice rails with scarf joints *(page 18)*.
◆ Fasten the rails in place with two nails at each end and at each stud mark in the same way as the stiles.
◆ Cut the remaining stiles to fit between the rails, then fasten each one *(left)*.

STILE / BASE

3. Adding the molding.

◆ With a power or manual miter saw, cut lengths of picture-frame molding to cover the edges of the rails and stiles, mitering the ends at 45 degrees.
◆ Fasten the molding in place—first drilling pilot holes if it is hardwood—by driving 1-inch finishing nails every 8 to 10 inches, starting with a stile *(right)* and working around the remaining three sides of the rectangle.
◆ Add trim to the remaining rectangles in the same way.

4. Adding raised panels.

◆ From $\frac{1}{2}$-inch plywood, cut a panel to fit inside each rectangle, making it 2 inches smaller all around than the inside dimensions of the rectangle.
◆ Spread panel adhesive *(page 32)* on the base piece, then center the raised panel in the rectangle, level it, and attach it with $2\frac{1}{2}$-inch finishing nails spaced every 8 to 10 inches around its perimeter *(left)*.
◆ Cover the edges of the raised panel with picture-frame molding.
◆ Fasten a $\frac{1}{2}$-inch-thick baseboard backer strip—scrap plywood will do—to the bottom of the base with 2-inch nails, then install the baseboards *(pages 12-18)*.
◆ Top off the wainscoting by adding a cap *(page 30)*.
◆ With a nail set, sink all the nails.
◆ If you will be painting the wainscoting, cover the nail holes with spackling compound and fill any gaps at the joints with caulk; otherwise, apply wood putty to hide the nail holes.

SIMULATING RAISED PANELS

Instead of adding raised panels as described above, you can simulate the look with a rectangle of molding. First, mark an outline within each rectangle 2 inches smaller all around than the rectangle formed by the rails and stiles. Then, miter lengths of molding such as chair rail to frame the outlines. Nail the molding to the plywood base with 1-inch finishing nails.

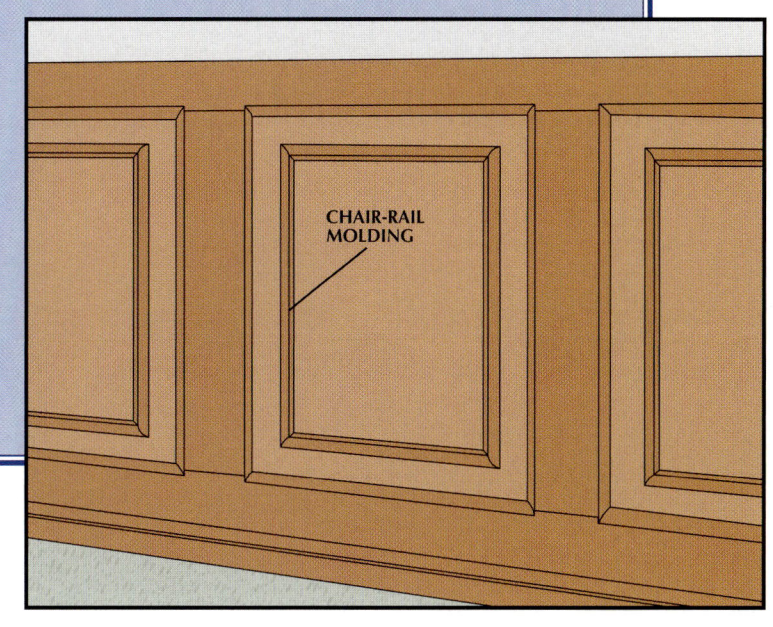

CHAIR-RAIL MOLDING

Creating a Fireplace Surround

Trimmed to match the other moldings in the house, a full surround can transform a plain fireplace into the focal point of a room.

Materials: Designed for a projecting brick fireplace, the unit shown on these pages is intended to be painted. The frame and trim are made of medium-density fiberboard (MDF), and the screws are counterbored and covered with spackling compound. If you prefer a stained surround, build it of A-C-grade softwood plywood or A-2 hardwood plywood and solid-wood molding. Counterbore the screws and plug them, and cover the exposed ends of the panels with edge banding.

For a more ornate look, replace the decorative trim with pilasters, such as the ones in the entryway on pages 80 to 89, or with simulated panels *(pages 33-35)*. You can also incorporate prefabricated carved details into the design.

TOOLS

Table saw
Straightedge
C-clamps
Sawhorse
Doweling jig
Electric drill with combination bit and masonry bit
Rubber mallet
Pipe clamps
Handscrew clamps
Screwdriver
Pry bar
Backsaw
Carpenter's level
Chalk line
Router
Power or manual miter saw
Nail set

MATERIALS

Medium-density fiberboard ($\frac{3}{4}$", 1")
Crown molding ($2\frac{1}{2}$")
Chair-rail molding
Picture-frame trim
Shoe molding
Shims
Fluted dowels ($\frac{3}{8}$")
Flooring screws (2")
Masonry screws (2")
Finishing nails ($1\frac{1}{4}$", $1\frac{1}{2}$")
Wood glue
Construction adhesive
Spackling compound

SAFETY TIPS

Protect your eyes with goggles when hammering or drilling, or when using a power saw.

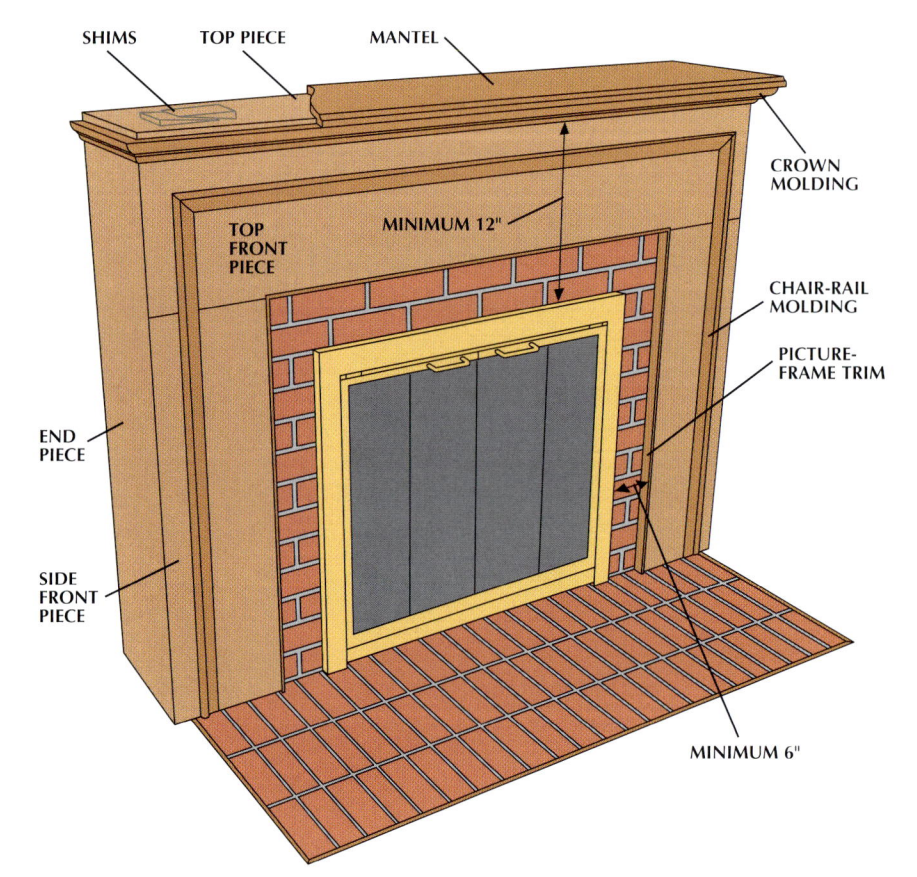

Anatomy of a surround.

The surround at left is a simple four-sided box. Its front face consists of two side pieces joined to a top piece with dowels. These sections can be as wide as desired as long as they are kept far enough away from the firebox to meet code requirements—generally 6 inches, but 12 inches for any elements, such as the mantel, that project outward more than $1\frac{1}{2}$ inches. The front is at least 2 inches wider than the existing fireplace—to overlap the end pieces and to allow for irregularities in the brickwork—and $\frac{1}{2}$ inch taller to accommodate the shims that keep the assembly level. End and top pieces complete the box, and a mantel is glued in place and trimmed with crown molding. In this example, the front is decorated with chair-rail molding, and the rough edges of the opening are concealed with picture-frame trim.

ASSEMBLING THE PIECES

1. Preparing the front pieces.
◆ From ¾-inch MDF, cut side and top front pieces to the desired size.
◆ Assemble the pieces on a work surface. Draw a triangle or X across the joints as a guide to ensure that the pieces can be reassembled in the same position. Across the top front piece and a side front piece, use a straightedge to mark lines for dowel locations 2 inches from each edge *(right)*. Mark the other side front piece in the same way.
◆ Clamp a side front piece to a sawhorse. Align a doweling jig with the marks, then drill holes for ⅜-inch dowels slightly deeper than ½ inch in the edge of the piece *(inset)*.
◆ Drill dowel holes at the marks in the other side front piece and the top front piece in the same way.

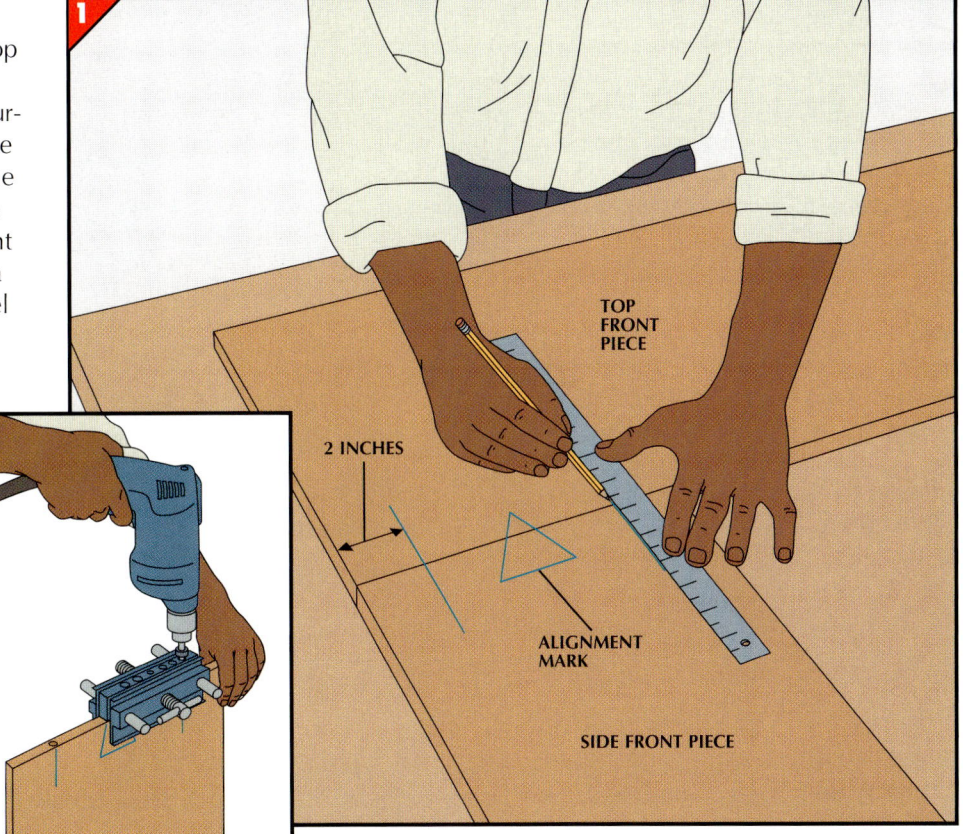

2. Joining the front pieces.
◆ Dab wood glue into the holes in the side front pieces and, with a rubber mallet, tap a 1-inch-long ⅜-inch fluted dowel *(photograph)* into each hole.
◆ Place the pieces on a work surface, laying a piece of wax paper under each joint.
◆ Insert glue in the matching dowel holes in the top front piece. Also spread glue on the contacting edges of the top front and side front pieces.
◆ Line up the dowels with the holes in the top front piece.
◆ With a pipe clamp, draw one joint together, protecting the pieces with wood blocks. Tighten the clamp until the joint is snug *(above)*.
◆ Join and clamp the opposite side of the front assembly in the same way.
◆ Clamp the assembly to the work surface with C-clamps to keep it flat.

3. Adding the end pieces.

◆ Cut two end pieces $\frac{1}{8}$ inch wider than the depth of the fireplace and as long as the height of the front assembly.
◆ Support each end piece with a handscrew clamp, then rest the front assembly on the end pieces.
◆ Fit an electric drill with a combination bit *(photograph)*, then drill shallow counterbored pilot holes for 2-inch flooring screws through the front pieces into one of the end pieces, locating the holes 2 inches from each edge and every foot in between *(right)*.
◆ Fasten the front to the end piece with wood glue and screws.
◆ Drill and fasten the other end piece in the same way.

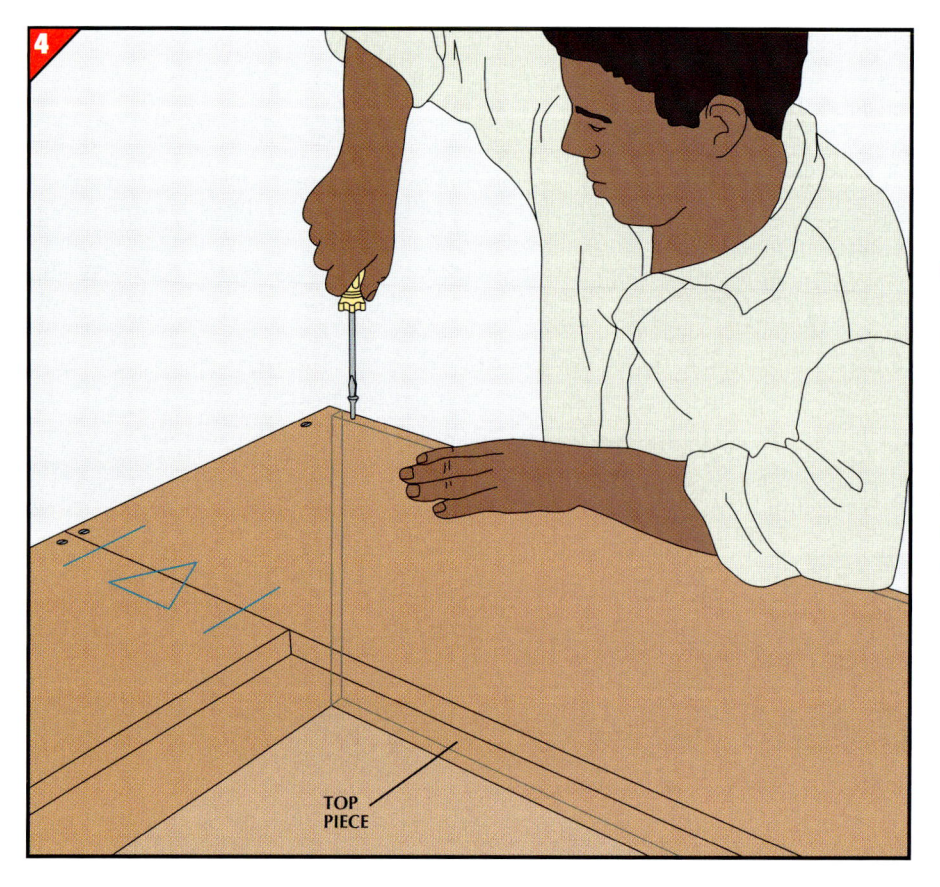

4. Fitting in the top piece.

◆ Cut a top piece the same width as the end pieces and long enough to fit between them.
◆ Slip the piece into position between the end pieces, making sure the edges of all the pieces are flush.
◆ Drill pilot holes and fasten the top piece to the top front piece and the end pieces with glue and screws *(left)*.

INSTALLING THE UNIT

1. Trimming the baseboards.
◆ With a pry bar, gently remove any shoe molding on both sides of the fireplace.
◆ Set the surround in place and mark a line on each side where it touches the baseboard *(right)*.
◆ Remove the unit and trim the baseboard at the lines on each side with a backsaw; pry the waste sections of baseboard away from the wall.

SHOE MOLDING

2. Shimming the unit.
◆ Reposition the surround, fitting it between the cut baseboards. Place a level on top of the unit and, if necessary, raise it with a pry bar and tap in shims until it is level *(left)*.
◆ Mark a line where the top of the unit meets the wall, then remove the unit and any shims from under it.
◆ Snap a chalk line on the wall $\frac{3}{4}$ inch below the first marked line to locate the bottom surface of the top board.
◆ Set two pairs of shims on the fireplace, adjusting them until the top of each pair meets the lower of the two lines on the wall *(inset)*. Fasten the shims to each other and to the fireplace with construction adhesive.
◆ Mark a vertical line on the wall above each pair of overlapping shims in line with the center of a brick.

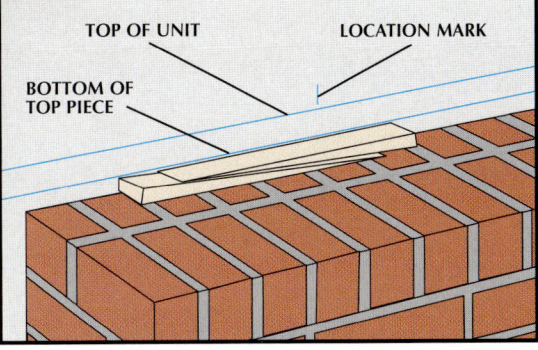

TOP OF UNIT   LOCATION MARK
BOTTOM OF TOP PIECE

3. Fastening the surround.
◆ Place the unit against the fireplace.
◆ Directly in front of each location mark, drill a shallow counterbored hole for a 2-inch masonry screw in the top of the unit.
◆ Drill similar holes through the front of the surround as well, positioning one on each side of the firebox opening 2 inches from the bottom of the unit and near the middle of a brick.
◆ With a masonry bit sized to the masonry screws *(photograph)*, drill a pilot hole through each counterbored hole.
◆ Screw the surround in place *(right)*.

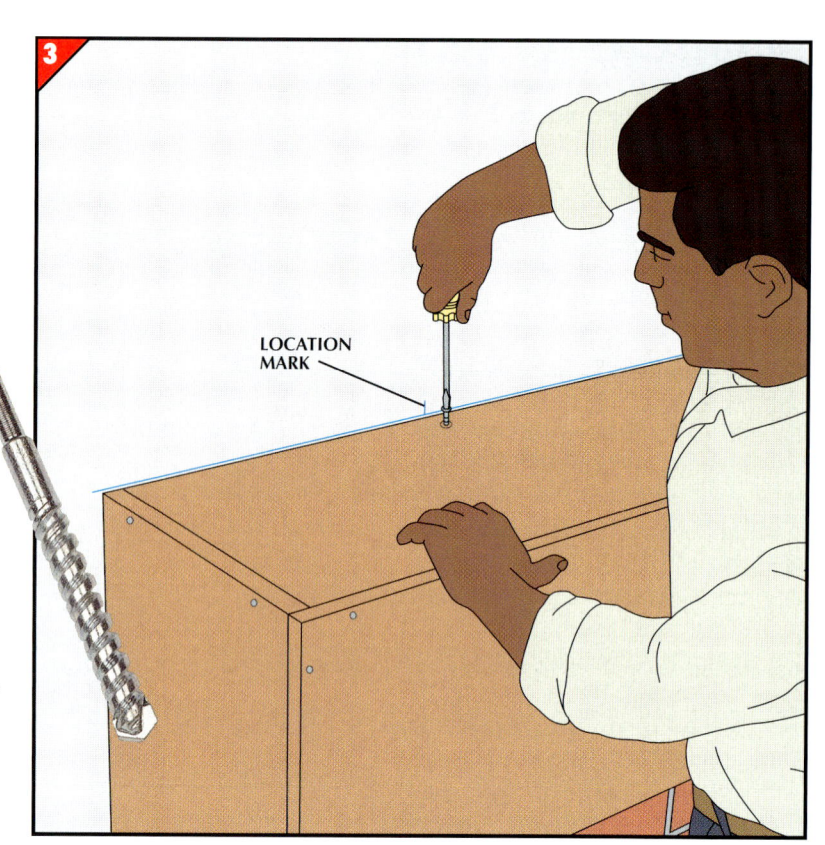

4. Adding the mantel.
◆ Measure the projection of $2\frac{1}{2}$-inch crown molding *(page 20)* and add $\frac{1}{2}$ inch to this figure.
◆ Cut a mantel from 1-inch MDF, sizing it to overhang the top of the unit at the front and both ends by the calculated measurement. If desired, shape the front edge and ends with a router *(page 11)*.
◆ Draw a line on the wall at the center of the surround and mark the center of the mantel on its back edge.
◆ Apply construction adhesive to the top of the surround.
◆ Set the mantel in place with the center marks aligned *(above)*.

ADDING THE TRIM

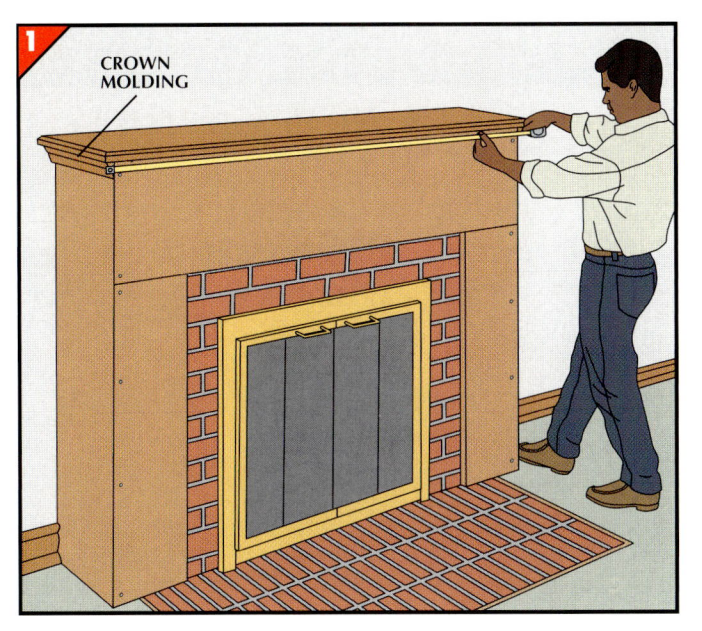

1. Installing crown molding.
◆ Trim a piece of crown molding for one side of the mantel with a power or manual miter saw, making a compound cut at the front end *(page 21)* so the short edge matches the depth of the fireplace surround. Cut a second piece for the other end of the unit.
◆ Fasten the crown molding to the mantel and the ends of the unit with wood glue and a $1\frac{1}{2}$-inch finishing nail at each end.
◆ Measure the distance between the two pieces of molding already installed *(above)*, and cut a length of crown to fit. Fasten it to the mantel and the front of the unit with glue and nails every 8 to 10 inches.
◆ Nail through each mitered corner from both sides of the joint.

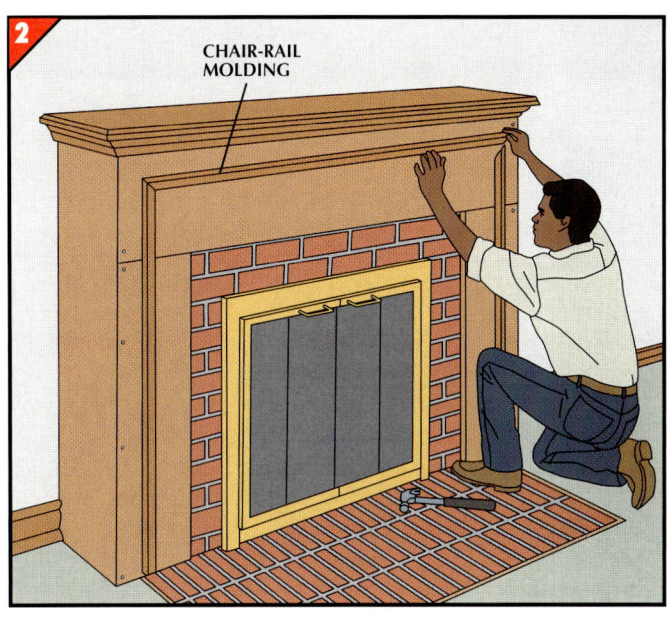

2. Adding decorative trim.
◆ Mark lines for the outside edges of the trim about one-third of the way from the top and sides of the surround.
◆ Measure from the floor to the horizontal line and miter two lengths of chair-rail molding at 45 degrees to fit.
◆ Align the outside edge of one piece of trim with one of the vertical lines and install it with wood glue and $1\frac{1}{4}$-inch finishing nails spaced every 8 to 10 inches. Install the other piece of vertical trim in the same way.
◆ Measure between the points of the vertical trim and cut another length to fit between them horizontally.
◆ Set the horizontal piece in place and check the fit *(above)*; if necessary, correct the miters *(pages 51-52)*.
◆ Fasten the horizontal trim with glue and nails.

3. Hiding the edges.
◆ Miter one end of a length of picture-frame molding *(inset)*. Measure the height of the inside edge of the opening in the surround and cut off the end of the molding to fit. Cut a piece of molding for the other side of the opening.
◆ Install the vertical pieces of molding with wood glue and nails spaced every 8 to 10 inches *(right)*.
◆ Measure between the points of the vertical molding and miter both ends of a third piece to fit. Apply glue to this piece, flex it to snap it into place, and add nails every 8 to 10 inches.
◆ With a nail set, sink the exposed nails on all the pieces of trim. Cover all screw and nailheads with spackling compound.
◆ Trim and replace the shoe molding on each side of the fireplace. Add shoe molding *(page 19)* along the bottom edges of the surround, if desired.

2 Windows and Trim

Installing modern factory-made windows can improve the appearance of your home as well as increase its energy efficiency. Depending on your decorative tastes, you can choose either picture-frame or stool-and-apron casing to cover the gap between the window frame and the interior wall framing.

Installing a Window 44

Fitting a Double-Hung Unit

Picture-Frame Casing 46

Extending the Jambs
Fastening the Trim
Correcting Imperfect Miters

Stool-and-Apron Trim 53

Installing the Casing

Marking a reveal line on the jambs →

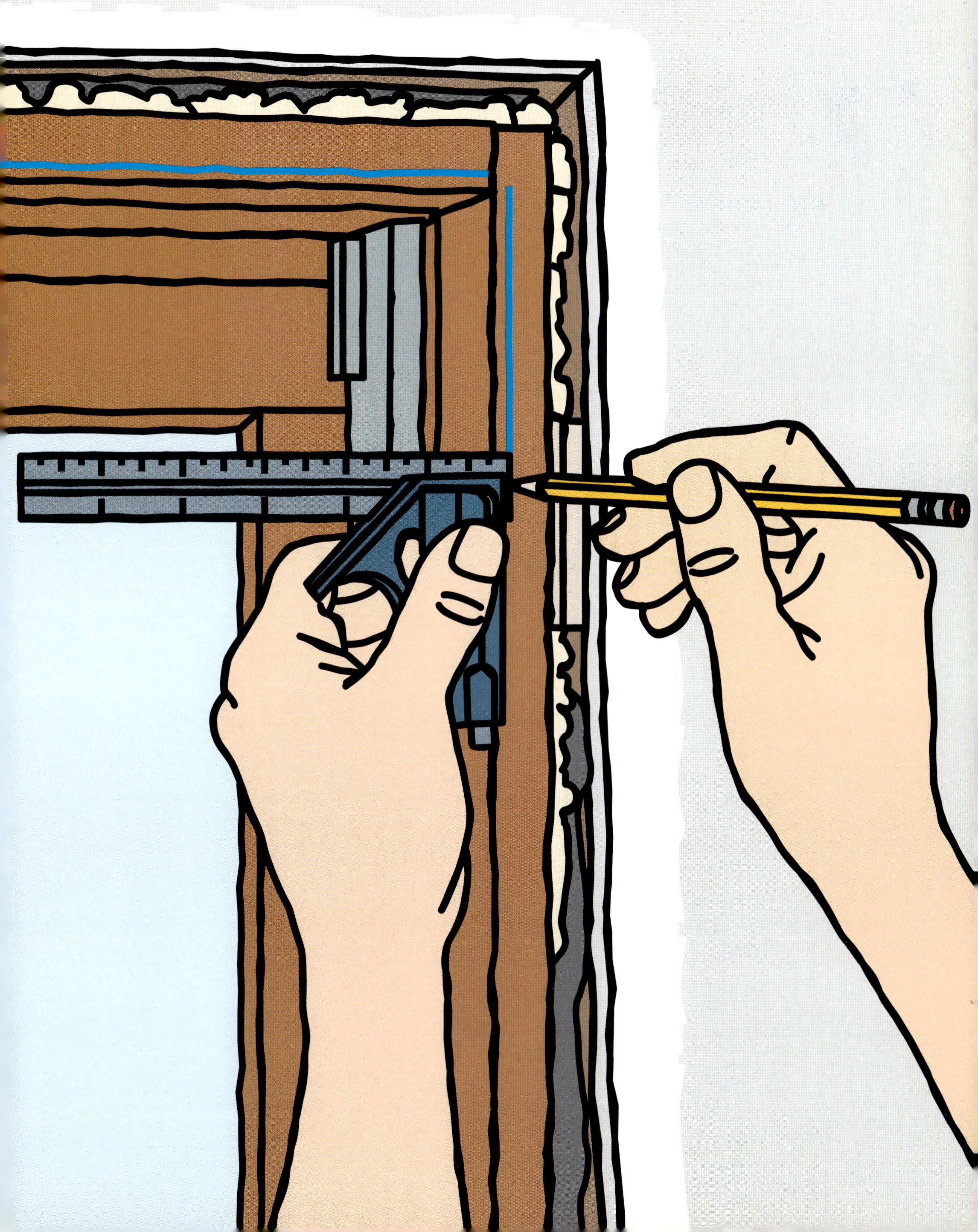

Installing a Window

Although it is possible to replace damaged sashes in a window with a sound frame, in some cases you will need to install a new window. Factory-made windows are generally prehung—they include both the sashes that hold the panes of glass and the jambs that make up the window frame.

Choosing a Window: Prehung windows are available in a variety of styles. The most common is the traditional double-hung window in which two sashes slide up and down within the jambs *(below)*. When buying a window, consider its insulating properties, which can be affected by the jamb material and kind of glass in the unit.

Purchase a window that is about $\frac{1}{2}$ inch smaller on all sides than its rough opening to allow for shims to level the unit and for insulation. (If you are framing a new rough opening, choose the window first, then size the opening to accommodate the window.)

Installing the Unit: To install a wood-frame window, nail the jambs to the rough framing through wood shims *(opposite)*; aluminum- and vinyl-frame windows are generally fastened to the outside of the house through a nailing flange around the perimeter of the window. Once the unit is secured in place, insulate the spaces around the jambs with fiberglass or, preferably, a low-expanding foam.

TOOLS
Hammer
Carpenter's level
Spring clamps
Utility knife
Nail set
Putty knife

MATERIALS
Drip cap
Finishing nails ($3\frac{1}{2}$")
Galvanized finishing nails ($3\frac{1}{4}$")
Shims
Wood putty or spackling compound
Expanding-foam insulation

SAFETY TIPS

Goggles protect your eyes when you are hammering. Wear work gloves when handling aluminum drip cap and rubber gloves when working with insulation.

FITTING A DOUBLE-HUNG UNIT

1. Positioning the window.
◆ For a new opening in a wall with wood or vinyl siding, buy a prefabricated drip cap and slip it into place between the siding and the building paper *(inset)*.
◆ Have a helper outside the house set the window in place so it is centered in the opening and the top brickmold fits up under the drip cap *(right)*.
◆ While you hold the window steady from inside, have the helper tack one corner of the top brickmold to the rough-opening header with a $3\frac{1}{4}$-inch galvanized finishing nail.

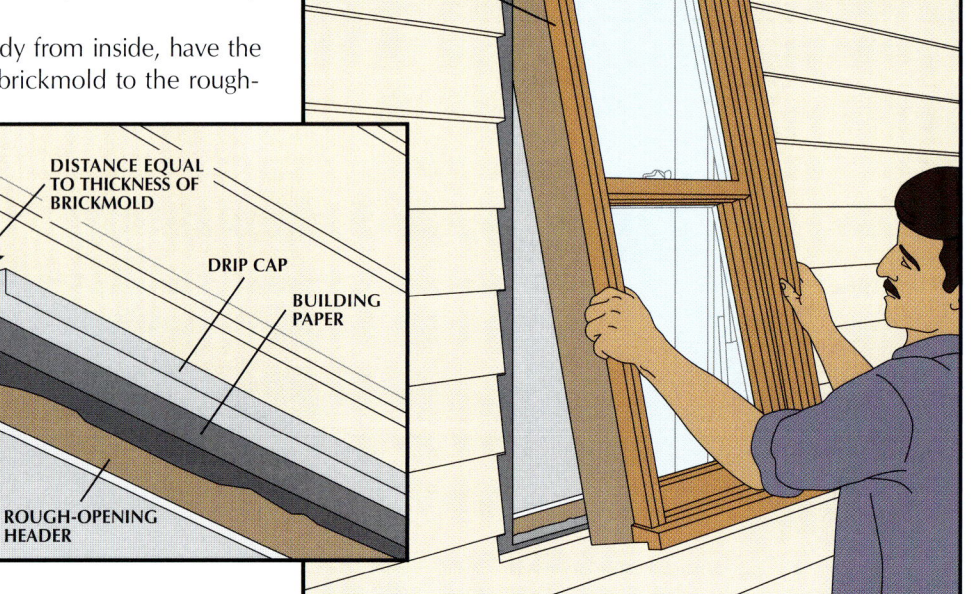

2. Leveling and centering the window.

◆ Inside, clamp a carpenter's level to the underside of the head jamb.
◆ While the helper steadies the window from outside, insert shims between the side jambs and studs at the top of the rough opening.
◆ Holding up one corner of the window, slip a shim between the window horn and the rough sill *(left)*. Add a shim under the other horn. Insert additional shims under the horns, as needed, to level the window.
◆ Have your helper nail the other top corner of the brickmold to the header.
◆ Insert shims between the side jambs and the studs at the middle and bottom of the window, being careful not to bow the jambs.

3. Nailing the window in place.

◆ Drive a $3\frac{1}{2}$-inch finishing nail through the side jamb and shims into the stud at each shim location *(above)*.
◆ Score the shims along the window jambs with a utility knife, then break them off flush with the jambs.
◆ Outside, nail the brickmold to the rough frame every 12 inches with $3\frac{1}{4}$-inch galvanized finishing nails.
◆ Pull the drip cap down tightly against the edge of the head brickmold.
◆ With a nail set, sink all the nails. If you plan to stain the trim, fill the holes with wood putty; otherwise, use spackling compound.

4. Insulating with foam.

Fill the space between the window jambs and the framing with expanding-foam insulation, working from bottom to top when filling the gaps around the sides of the window *(above)*. Use the product sparingly—too much of it can cause the jambs to bow inward when the foam expands.

Picture-Frame Casing

Window casing frames a window inside the house. The simplest form, called picture-frame casing, is made of four mitered pieces of casing stock. Moldings suitable for casing come in many styles. Unless you wish to call special attention to the windows, choose one $1\frac{1}{2}$ to $3\frac{1}{2}$ inches wide with modest contours.

Preparing for the Job: Picture-frame casing requires the front edges of the window jambs to be flush with the interior wall. If the jambs fall short of this goal by more than $\frac{1}{4}$ inch, install a jamb extension *(opposite)*; for a smaller difference, shave the wall flush with the jamb *(page 48)*. You'll also need to mark a reveal around the jambs or jamb extension so that the inside edges of the casing pieces can be offset from the inside of the jambs by a uniform amount all around the window. A reveal lends an attractive look to the window, and it is easier to fashion one than to install the casing flush with the jambs.

Fitting the Miters: The trickiest part of installing picture-frame casing is achieving neat miters. For best results, cut and install the head and side casings first, then fit the sill casing. Miters that do not fit perfectly can often be corrected *(pages 51-52)*. Casings can be painted to match or contrast with the wall color, but they also may be stained and varnished.

TOOLS
Table saw
Electric drill
Screwdriver
Jack plane
Hammer
Utility knife
Nail set
Block plane
Straightedge
Rasp
Combination square
Power or manual
  miter saw
Wood chisel

MATERIALS
Lumber for jamb
  extension
Shims
Casing stock
Wood glue
Wood screws
  (2" No. 8)
Finishing nails
  ($1\frac{1}{2}$", 2")
Expanding-foam
  insulation
Wood putty or
  spackling
  compound

SAFETY TIPS
Protect your eyes with goggles when hammering, and put on rubber gloves before applying expanding-foam insulation.

A picture-framed window.
Four pieces of casing surround this window—head casing along the top, sill casing along the bottom, and side casing. The pieces meet at the corners with 45-degree miters, and are offset from the jambs or jamb extension by $\frac{1}{4}$ inch, leaving a reveal. The trim is nailed to both the window jambs or extension and the rough framing.

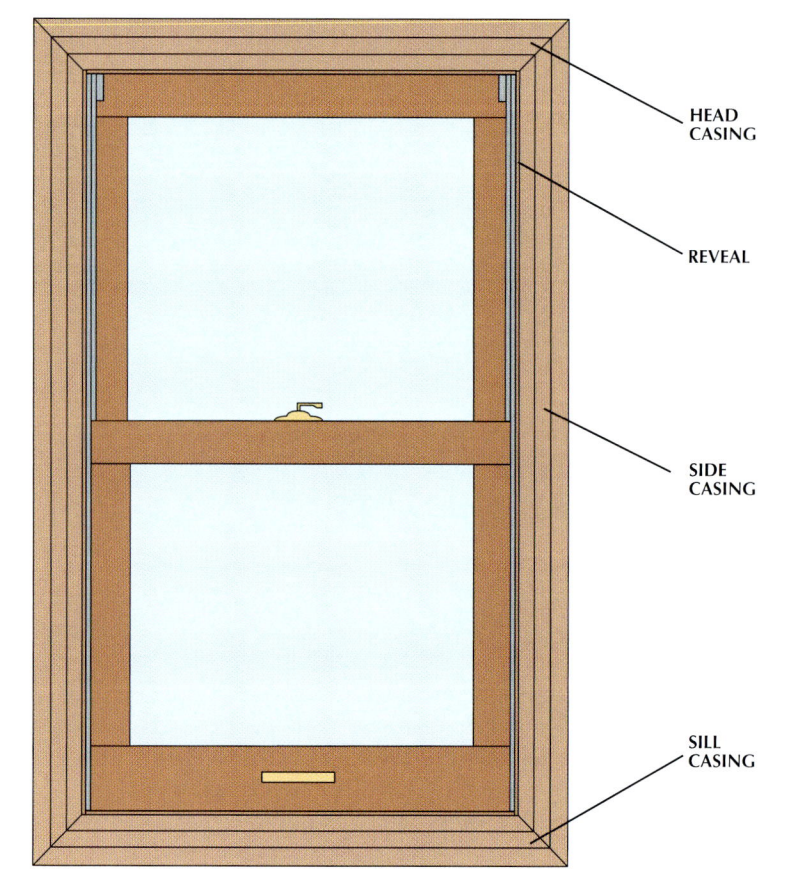

- HEAD CASING
- REVEAL
- SIDE CASING
- SILL CASING

EXTENDING THE JAMBS

1. Making an extension.
◆ Measure the distance between the front edge of the window jambs and the inside wall *(inset)*.
◆ On a table saw, rip lumber of the same thickness as the jamb to match this measurement.
◆ Cut pieces of this stock to make a frame having interior dimensions $\frac{1}{4}$ inch longer and wider than the inside measurements of the jambs.
◆ Assemble the frame with the sawn edges facing up; drill pilot holes for two 2-inch No. 8 wood screws near each end of the side pieces and into the top and bottom pieces *(right)*, then drive the screws.

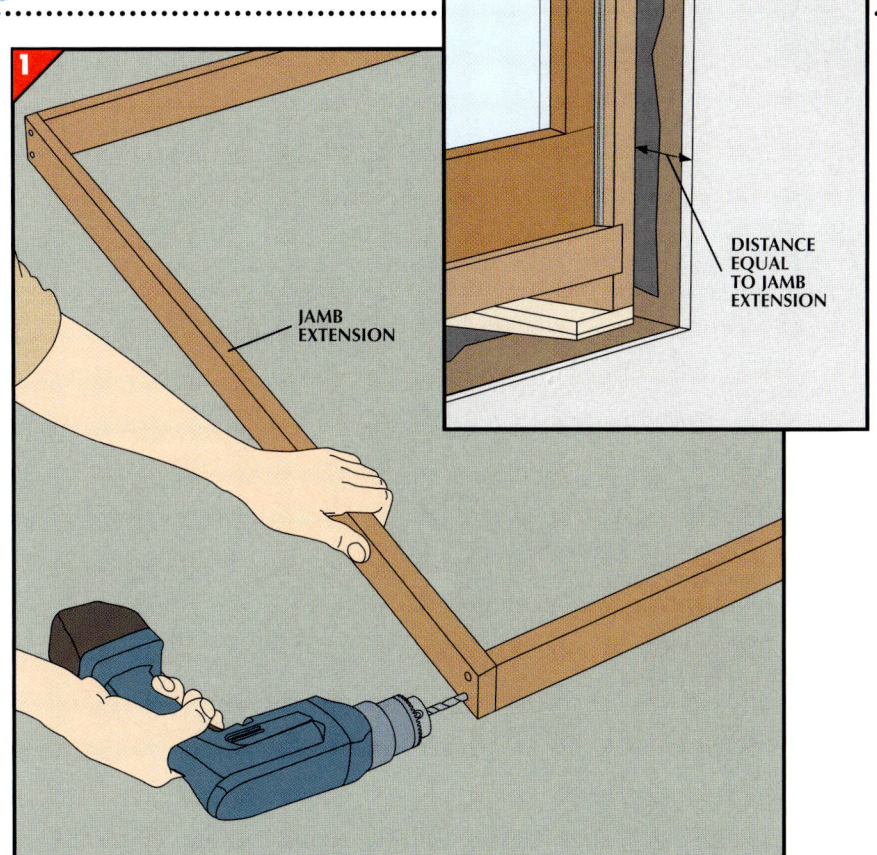

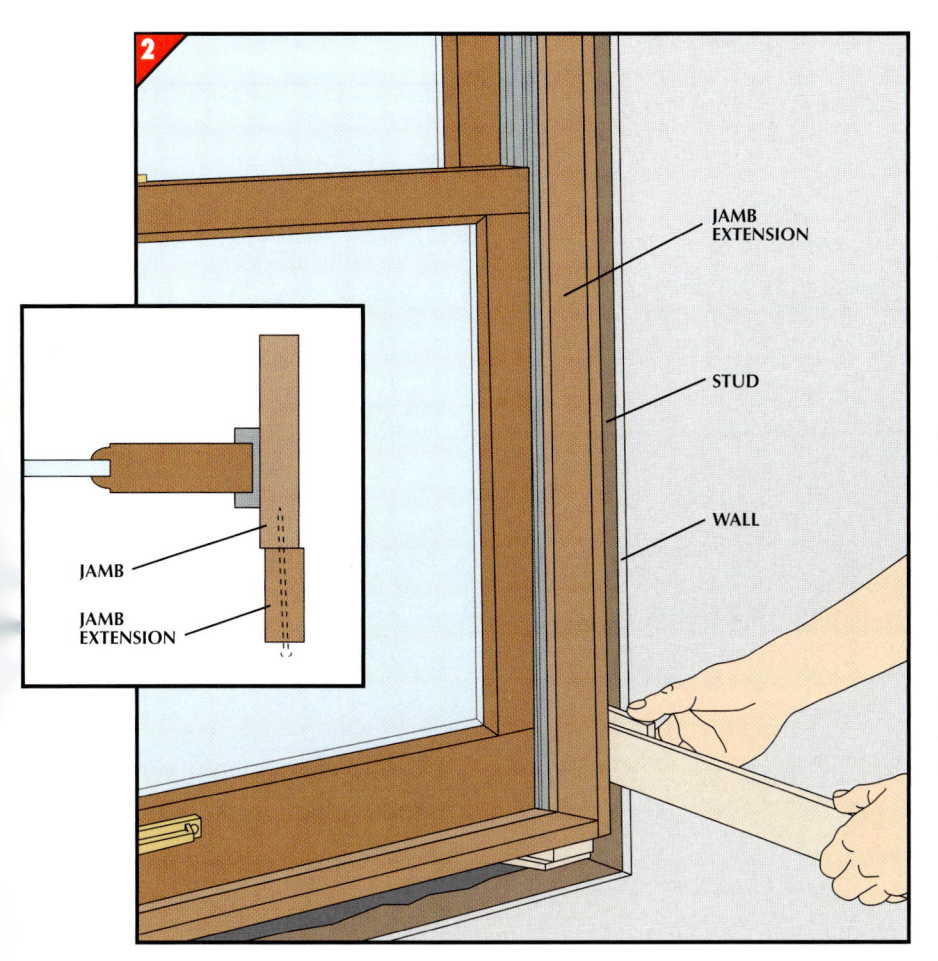

2. Installing the extension.
◆ Position the jamb extension against the jambs with the sawn edges out, leaving a $\frac{1}{4}$-inch reveal all the way around *(inset)*. If the extension sticks out beyond the wall surface, remove it, and shave the edges flush with a jack plane.
◆ Place shims on the rough sill, spread wood glue on the inside edges of the extension, and position it on the jambs. Slide shims between the studs and the extension *(left)*, as when installing a window *(page 45, Step 2)*.
◆ Into the outside edges of the extension, drill pilot holes every 8 inches for finishing nails long enough to penetrate the jambs by $\frac{3}{4}$ inch; then drive the nails.
◆ Score the shims with a utility knife and break them off.
◆ Sink the nails with a nail set.
◆ Add expanding-foam insulation around the extension *(page 45, Step 4)*.

FASTENING THE TRIM

1. Preparing to install the trim.

◆ If a jamb extension was not required, but the jamb is proud of the wall surface, shave it flush with a block plane. If the jamb is set back from the wall by $\frac{1}{4}$ inch or less, first use a straightedge and utility knife to cut through the wallboard paper slightly in from where the edge of the casing will be. With a rasp or forming tool *(photograph)*, shave the wallboard within the cutting line until it is flush with the jamb.

◆ Adjust a combination square to $\frac{1}{4}$ inch and butt its handle against the inside face of the jamb. Holding a pencil flush against the blade, slide the handle and pencil along the jamb to mark the reveal line *(right)*. Repeat the procedure to mark the other jambs.

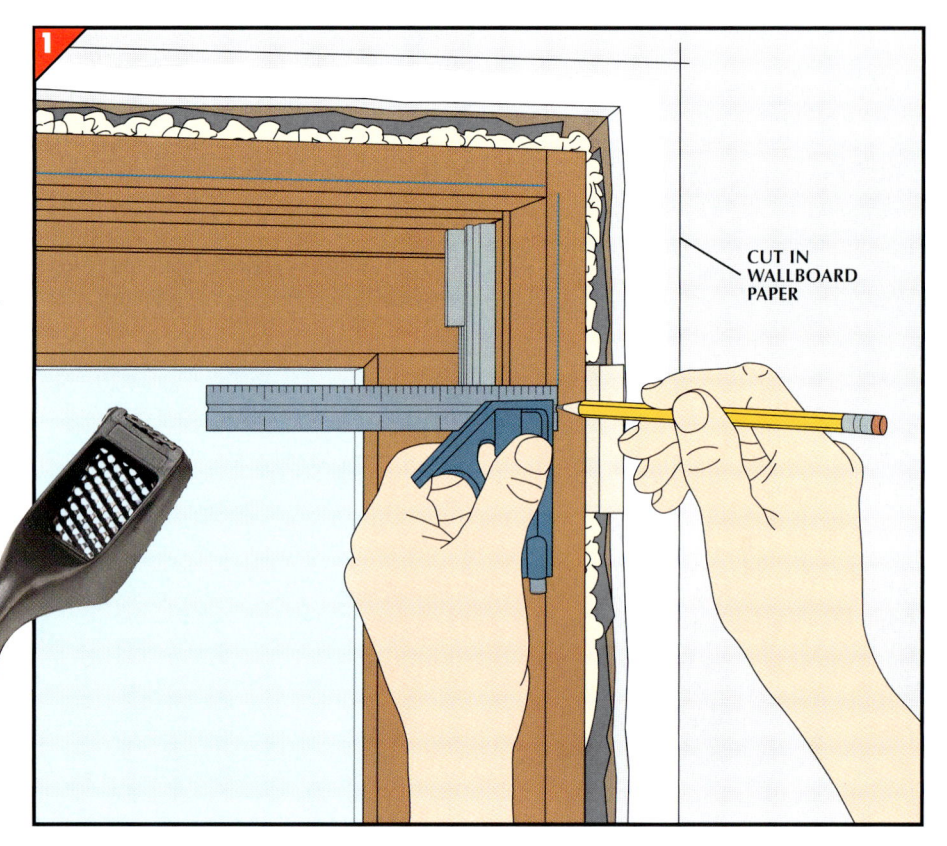

TRICKS OF THE TRADE

A Reveal Gauge

With the jig shown at right, you can mark a reveal for casing around window jambs with ease. Along one edge of a square piece of $\frac{3}{4}$-inch plywood or hardboard, saw a notch $\frac{1}{4}$ inch wide—the typical reveal for $\frac{3}{4}$-inch jamb stock. Cut a $\frac{1}{8}$-inch notch along another edge for use with jambs thinner than $\frac{3}{4}$ inch. Label the notches on the front and back of the gauge. To use the jig, butt the appropriate side against the jamb, then mark the reveal as described above for the combination square *(right, below)*.

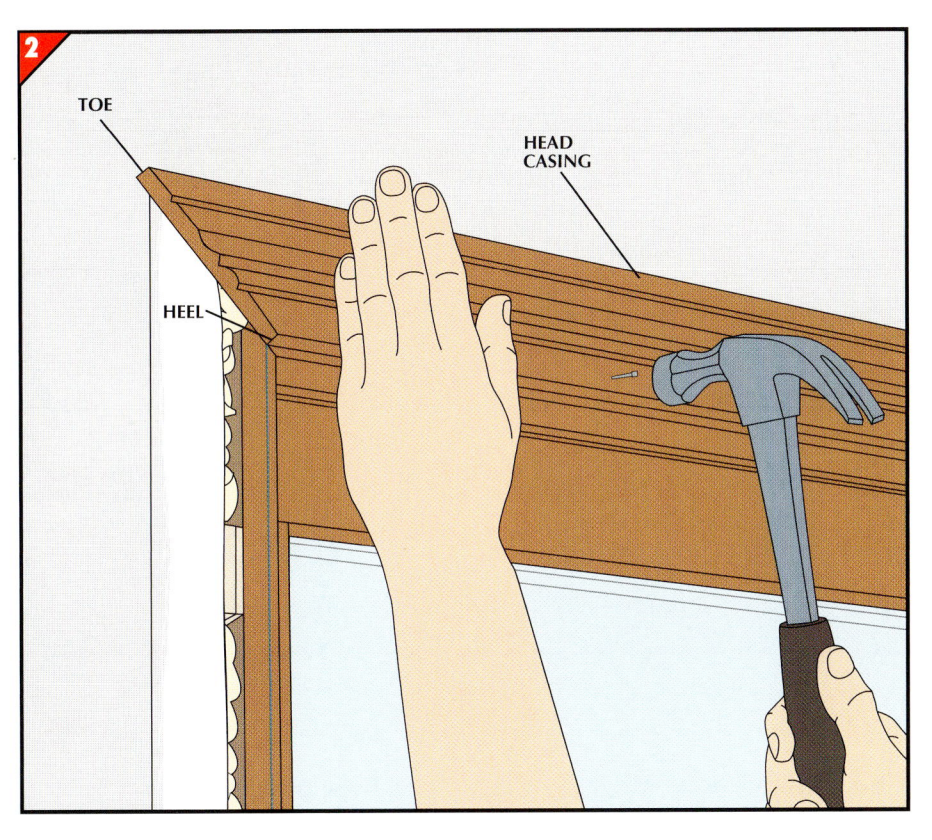

2. Installing the head casing.

◆ To determine the length of the head casing, measure the distance between the window jambs and add twice the reveal to your measurement.
◆ Miter both ends of the head casing at 45 degrees so the distance between the heels of the miters equals the length you calculated.
◆ Align the bottom edge of the casing with the reveal line. Then, fasten the trim with a pair of finishing nails every 6 inches—drilling pilot holes first if the casing is hardwood—driving a 1½-inch nail into the jamb (left) and a 2-inch nail directly above it through the wall and into the rough-opening header.

3. Nailing the side casing.

◆ Determine the length of the side casing pieces and miter their ends as for the head casing (Step 2).
◆ Set one piece in place; if the miter joint fits poorly, correct it (pages 51-52). Test and adjust the second piece of side casing in the same way.
◆ Spread a little wood glue on two of the miters that will be joined and set one piece of side casing in place.
◆ Starting at the top—drilling pilot holes first for hardwood—fasten the casing every 6 inches with a 1½-inch finishing nail driven into the jamb (right) and a 2-inch nail driven into the wall stud, stopping about 1 foot from the bottom.
◆ Fasten the second piece of side casing in the same way.
◆ If the face of one piece is proud of the other at either joint, carefully pare down the high spot with a sharp chisel.

4. Putting in the sill casing.

◆ Measure the distance between the side casings and miter both ends of the sill casing to fit.
◆ Test-fit the piece and correct the miters if necessary *(opposite and page 52)*.
◆ Spread a little glue on the miters and position the sill casing. Spacing nails 6 inches apart, fasten the casing to the jamb with $1\frac{1}{2}$-inch finishing nails *(above)* and to the rough sill with 2-inch nails—drilling pilot holes first if going into hardwood.
◆ Pull the bottom of the side casings tight against the sill casing and finish nailing them in place.
◆ If either of two adjoining pieces of casing is proud of the other, pare down the high spot with a sharp chisel.

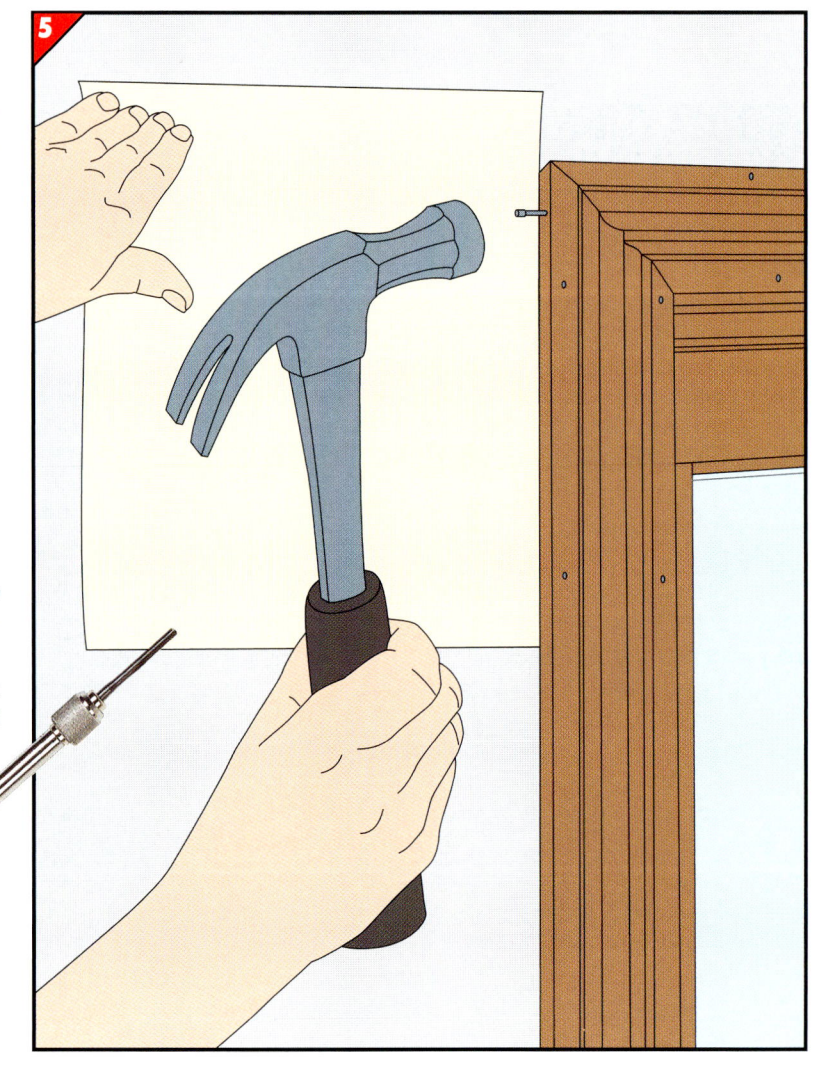

5. Cross-nailing the miter joints.

◆ Holding a piece of cardboard against the wall to protect the surface, drive a 2-inch nail through the side casing and into the end of the head casing *(right)*. If working with hardwood trim, bore a pilot hole first with a push drill *(photograph)*, which will fit into an area too tight for an electric drill.
◆ Cross-nail the remaining miters in the same way.
◆ Countersink nailheads with a nail set, then fill the holes with spackling compound—or with wood putty if you will be staining the trim.

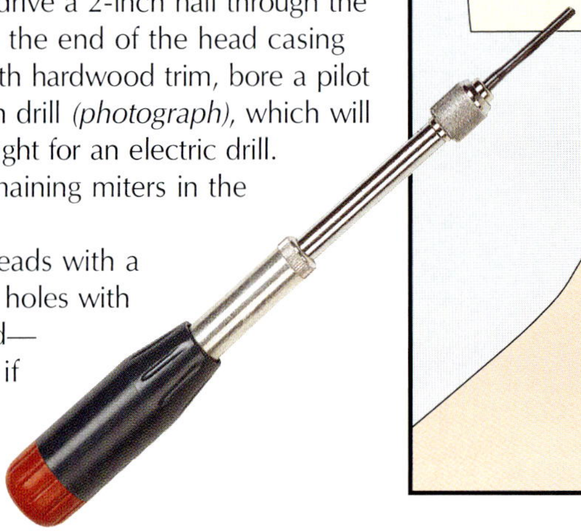

MODIFIED PICTURE-FRAME CASING

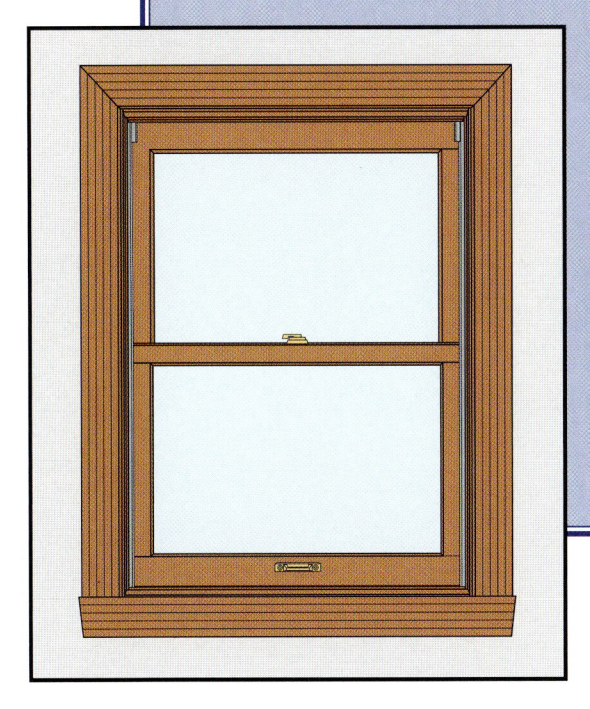

You can simplify the installation of picture-frame casing by using butt joints at the bottom instead of miters. Join the top and side casing with miters as for ordinary picture-frame casing *(pages 48-49, Steps 1-3)*, but cut the bottom ends of the side casing square. Then trim the ends of the sill casing to extend a little beyond the side pieces; for decorative effect, you can cut a shallow miter at each end of the sill casing. Then, simply butt the edge of the sill casing against the bottom ends of the side casing and nail it in place.

CORRECTING IMPERFECT MITERS

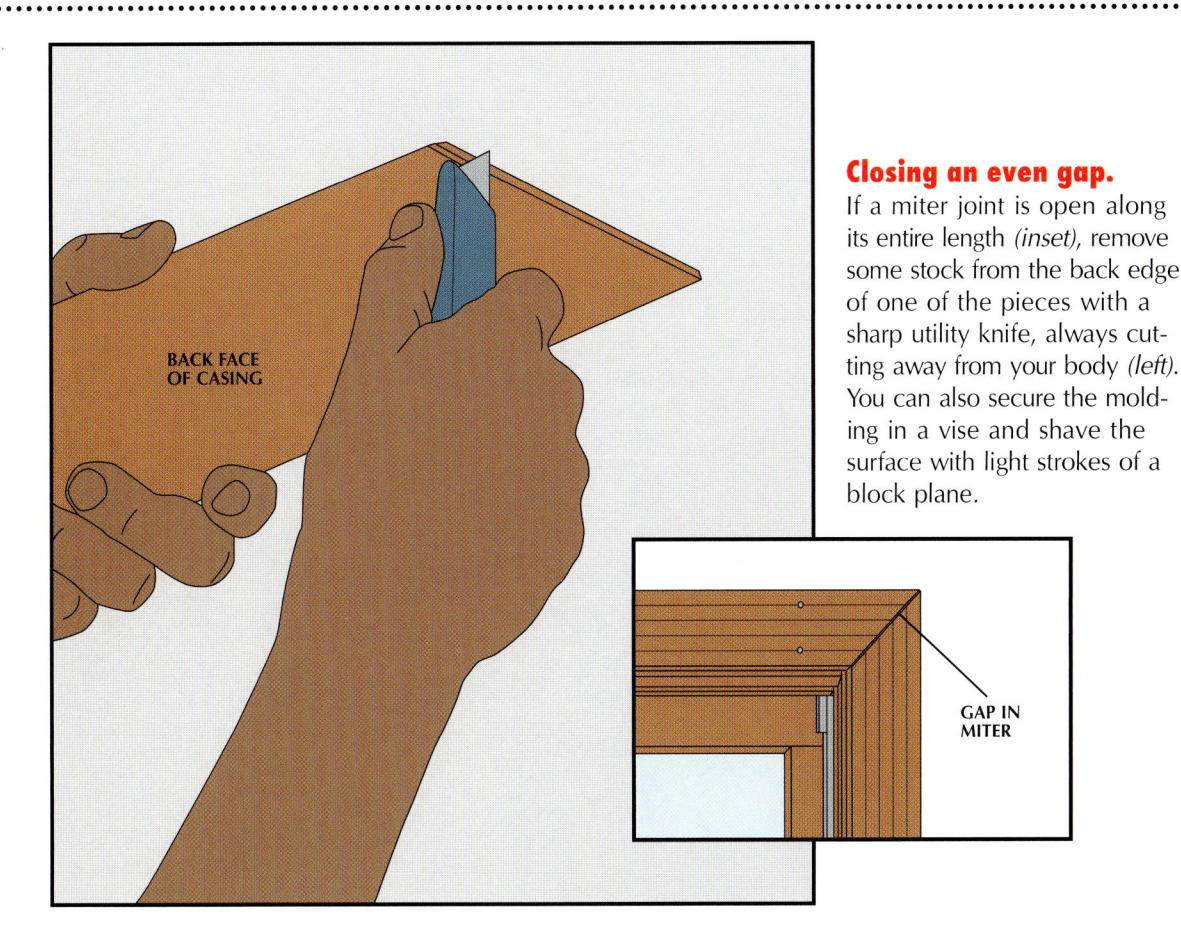

Closing an even gap.
If a miter joint is open along its entire length *(inset)*, remove some stock from the back edge of one of the pieces with a sharp utility knife, always cutting away from your body *(left)*. You can also secure the molding in a vise and shave the surface with light strokes of a block plane.

BACK FACE OF CASING

GAP IN MITER

Closing a gap at the heel.

You can eliminate a gap at the heel of a miter joint *(inset)* using a power miter saw.

◆ Attach a wood auxiliary fence to the regular fence of the saw and cut through the wood fence at 45 degrees. Swing the blade 45 degrees in the opposite direction and cut through the fence again.

◆ Set one of the casing pieces against the fence with the toe of the miter extending slightly beyond the fence, and slip a thin wedge between the casing and the fence 1 or 2 inches from the end of the board *(right)*.

◆ Cut the stock *(dashed line)*.

◆ Test-fit the joint and repeat the cut, if necessary, moving the wedge $\frac{1}{4}$ inch farther away from the end of the casing.

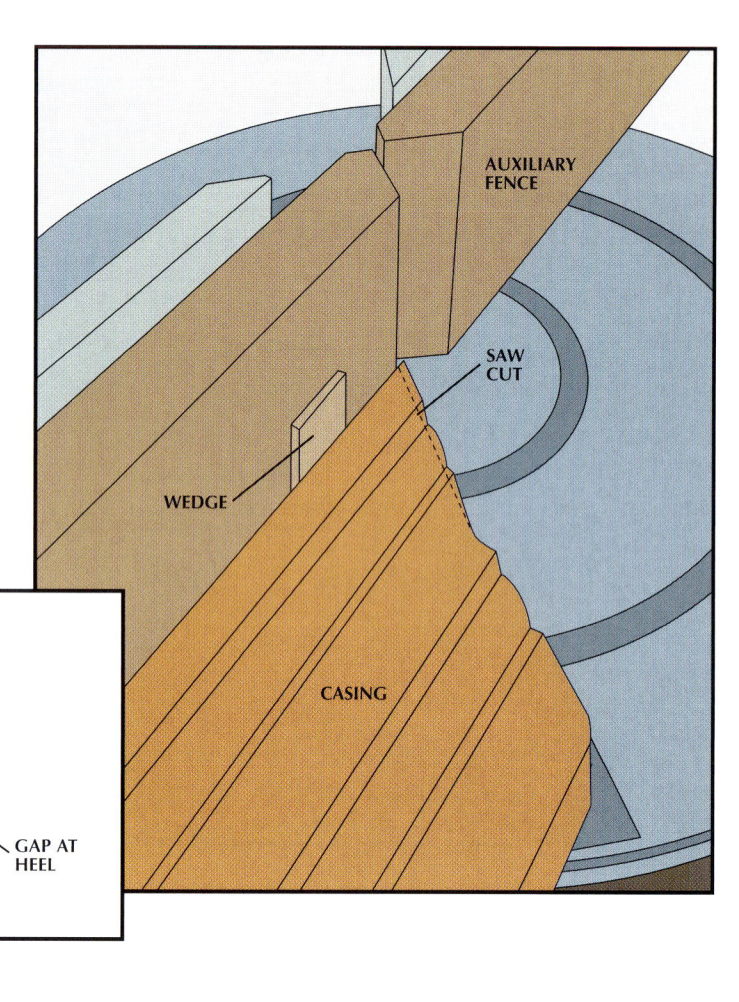

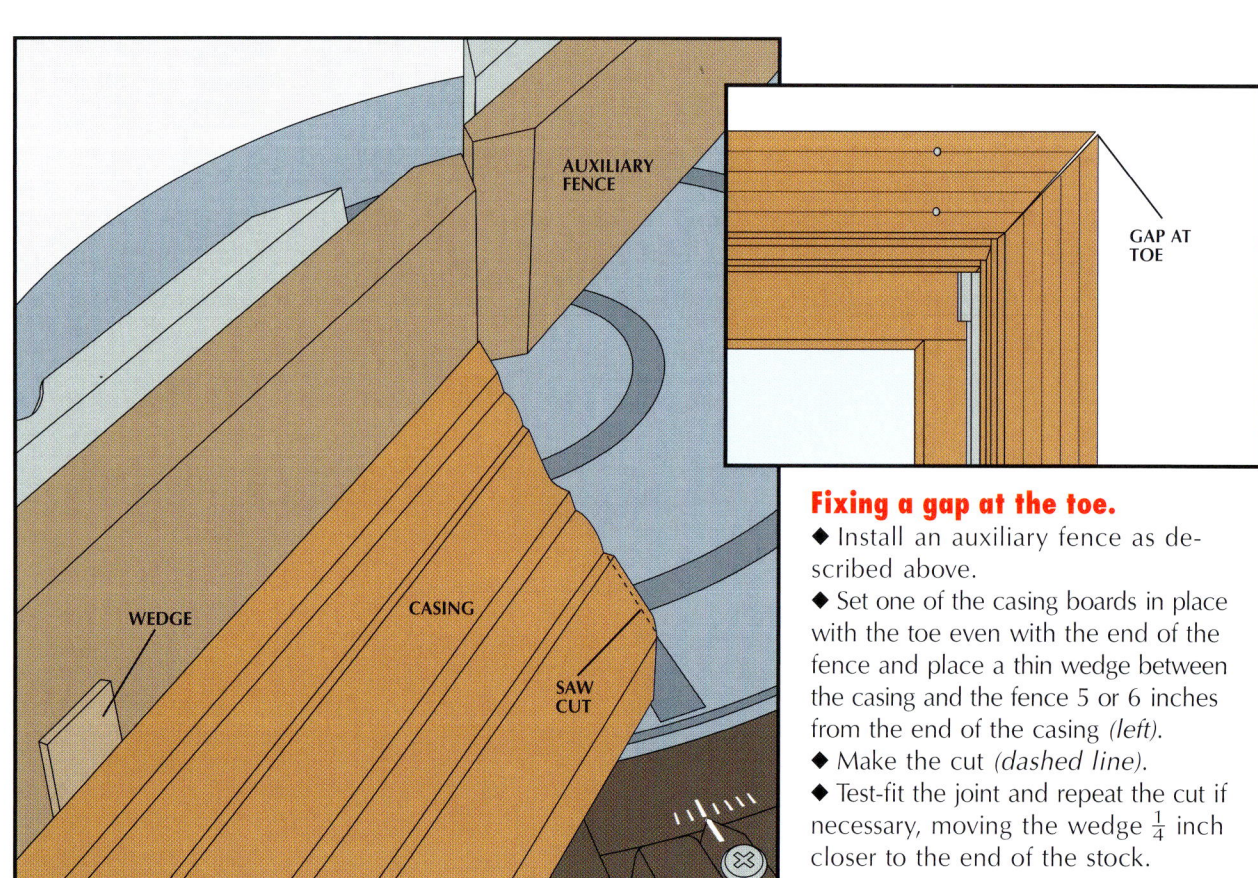

Fixing a gap at the toe.

◆ Install an auxiliary fence as described above.

◆ Set one of the casing boards in place with the toe even with the end of the fence and place a thin wedge between the casing and the fence 5 or 6 inches from the end of the casing *(left)*.

◆ Make the cut *(dashed line)*.

◆ Test-fit the joint and repeat the cut if necessary, moving the wedge $\frac{1}{4}$ inch closer to the end of the stock.

Stool-and-Apron Trim

Although more time-consuming to install than picture-frame casing, stool-and-apron trim lends a traditional look to a window. In this style, the head and side casings meet with butt joints instead of miters, the window sill is extended with a stool, and the gap under the stool is covered with an apron *(below)*.

Choosing the Trim: Even simple casings can look elegant with this style of trim, but wide or elaborate profiles are also appropriate. You can make the head casing slightly thicker than the side casing and, since the head and side casings meet with butt joints, you can even combine contrasting moldings.

Installation: On older windows, the sill often slopes toward the outside, requiring that the stool be custom cut to fit it over the sill. Modern windows such as the one shown here typically have a flat sill; in this case the stool can be simply joined to the sill with a butt joint *(page 55)*.

Once the stool is in place, the casing pieces are added. If the existing jambs sit below the surface of the inner wall, first add a modified jamb extension *(page 55, Step 4)*.

TOOLS

Router	Saber saw	Screwdriver
Straightedge	Utility knife	Carpenter's level
Combination square	Hammer	Block plane
Compass	Table saw	Rasp
	Power or manual miter saw	Nail set
	Electric drill	

MATERIALS

Stool and casing stock	Wood screws (2" No. 8)
Lumber for jamb extension	Finishing nails ($1\frac{1}{2}$", 2", $3\frac{1}{2}$")	Wood putty or spackling compound
Shims	Wood glue	

SAFETY TIPS

Protect your eyes with goggles when hammering or using a power tool.

A traditional casing design.

The window at right is trimmed in a stool-and-apron style. The stool extends off the sill, and is supported on shims and fastened to the sill with glue and nails. On each side of the casing, the stool is cut to leave two decorative horns that extend onto the wall—typically by $\frac{3}{4}$ inch.

The side and head casing are fastened to the jambs and wall framing, leaving a $\frac{1}{4}$-inch reveal around the edge of the jambs. The head casing extends past the side casing by the same amount as the stool horns. Rosettes at the top corners are an attractive alternative *(inset)*.

An apron with the same profile as the casing—or a compatible one—covers the gap between the underside of the stool and the wall. Fastened to the wall framing, the apron is beveled at each end and trimmed with mitered returns *(page 57)*.

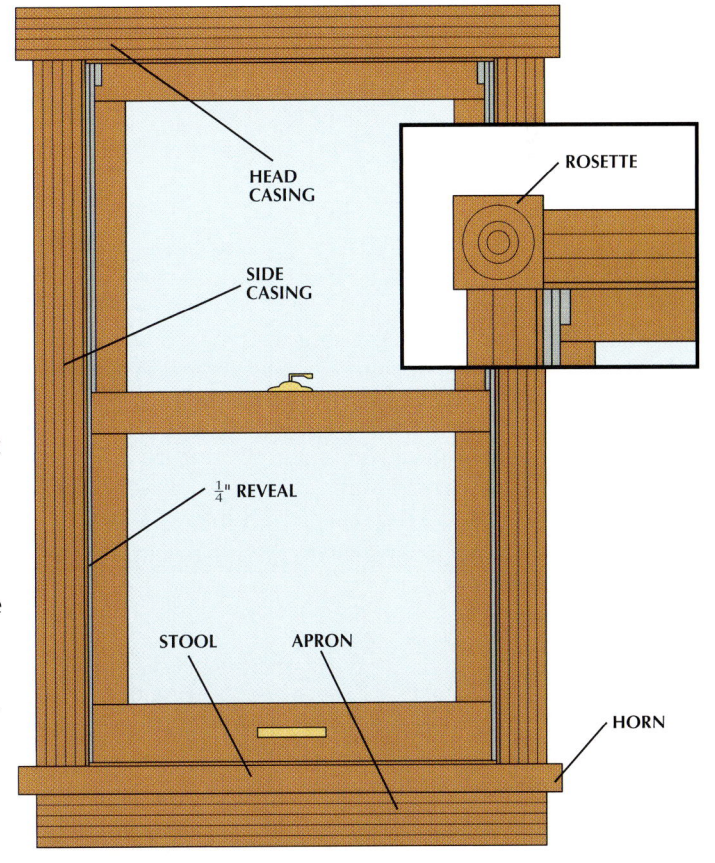

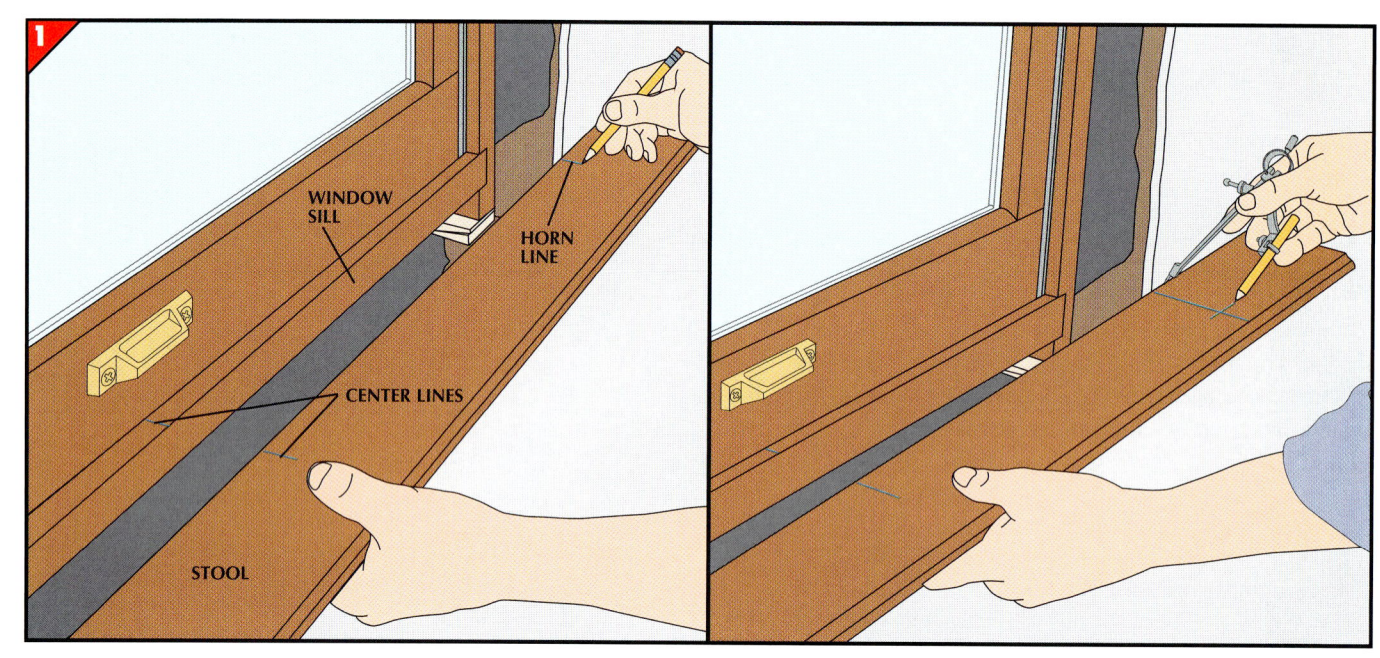

INSTALLING THE CASING

1. Marking the horns.
◆ To determine the required width of the stool, measure the distance between the window sill and the wall surface, then add the thickness of the casing you will use, plus $\frac{3}{4}$ inch. For the length, measure the width of the opening, then add twice the width of the casing, plus $1\frac{1}{2}$ inches.
◆ Cut a piece of stool stock to size, then use a router to shape the outside edge and ends of the stool, if desired.
◆ Mark the centers of both the stool and the sill.
◆ Hold the stool up to the wall and align the center marks with a straightedge, then mark the edge of the wallboard at each side of the window on the stool, indicating the inside edges of the horns *(above, left)*.
◆ With a combination square, extend both horn lines to the front edge of the stool.
◆ Adjust the legs of a compass to the gap at its widest point between the front edge of the window sill and the wallboard. With the stool in position against the wall, set the compass point at the edge of the wall and scribe a line for each horn *(above, right)*.

2. Trimming the stool.
◆ Keeping the same compass setting as in Step 1, mark a line along the length of the stool, running the compass point along the front edge of the window sill *(right)*.
◆ With a saber saw, cut out the horns as well as the waste strip from the inside edge of the stool.

3. Fastening the stool.

◆ Test-fit the stool, placing a set of shims under it at each end of the opening and in the middle to raise it flush with the window sill. Trim the shims flush with the wall.
◆ Apply wood glue to the contacting edges of the sill and stool.
◆ Repositioning the stool against the sill, drive a 3½-inch finishing nail down through the stool and the shims at one end into the rough framing *(right)*, then nail the other end of the stool in the same way.
◆ Let the glue dry. If the window jambs are not flush with the wall surface, install a jamb extension *(Step 4)*. Otherwise, install the side casing *(Step 5)*.

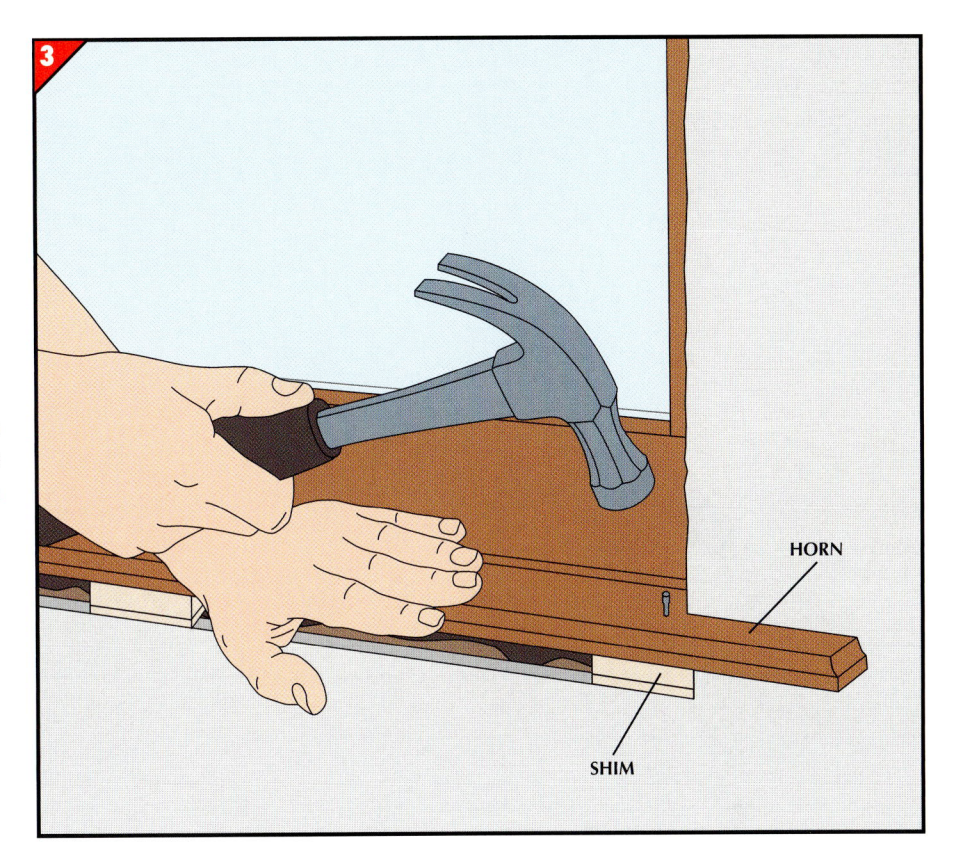

4. Extending the jambs.

◆ Build a standard jamb extension *(page 47, Step 1)*, but omit the bottom piece.
◆ Fit the frame against the window jambs, resting the bottom on the stool *(left)*. Shim the extension to center it in the opening, checking it for square and level with a carpenter's level.
◆ Drill pilot holes through the extension and into the jambs every 6 inches for finishing nails long enough to penetrate the jambs by ¾ inch.
◆ Nail the extension to the jambs.

5. Fastening the side casings.

◆ Prepare to install the trim, marking a $\frac{1}{4}$-inch reveal around the jamb or jamb extension *(page 48, Step 1)*.
◆ Measure from the stool to the reveal line on the top jamb, then cut the side casings to length with both ends square.
◆ Every 6 inches, nail the casings to the jambs with $1\frac{1}{2}$-inch finishing nails *(above)* and to the wall studs with 2-inch nails—drilling pilot holes first if going into hardwood.

6. Fitting the head casing.

◆ Cut the head casing to the same length as the stool, and center it over the side casing pieces.
◆ Drive $1\frac{1}{2}$-inch finishing nails into the head jamb *(above)* and 2-inch nails into the rough header, spacing the nails about 6 inches apart; for hardwood, drill pilot holes.

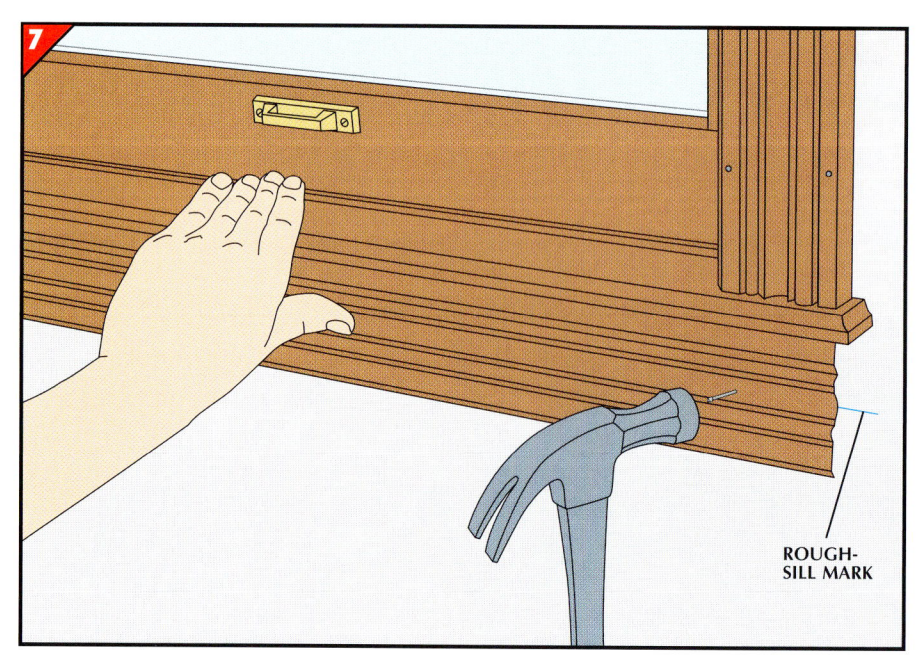

7. Fastening the apron.

◆ Measure the distance between the outside edges of the side casings and cut an apron to this length, beveling each end inward at 45 degrees.
◆ Mark the center of the rough sill's outer edge on the wall on each side of the opening.
◆ Center the apron under the window and butt it against the stool, then drive 2-inch nails every 6 inches at the level of the mark *(above)*—drilling pilot holes first if going into hardwood.
◆ With a nail set, sink all the finishing nails, then fill the holes with wood putty—if you plan to stain the casing—or with spackling compound.

TRICKS OF THE TRADE

Propping an Apron

To free up your hands when you are nailing a window apron in place, brace the apron with a board as shown at right. Cut a piece of thin scrap stock slightly longer than the distance between the apron and the floor, and wedge the prop tightly between the apron and the floor so it bows slightly.

BEVELED END

RETURN

8. Adding the apron returns.

◆ Make a mitered return by cutting a 45-degree bevel across the end of a piece of molding, then trimming off a narrow wedge *(inset)*.
◆ Apply wood glue to one end of the apron and the straight edge of the return, then press the return against the end of the apron *(left)* and secure it with masking tape.
◆ Cut and attach a return to the other end of the apron in the same way.
◆ Remove the masking tape once the glue has dried.

3 Hanging Doors

Whether you choose to install a ready-made door or build one yourself, the job requires good carpentry skills. A properly constructed unit must swing easily, close securely, and be finished off with a well-fitting casing. For an exterior door, you can take these skills one step further and create a formal entryway to dress up the house.

Building a Frame-and-Panel Door — 60
Grooving the Frame Parts
Assembling the Door

Installing Jambs — 65
Making and Fastening the Frame
Adding Stop Molding

Hanging the Door — 69
Attaching Hinges

Installing a Lockset — 73
Putting in a Privacy Lock

Finishing Up with Trim — 77
Attaching the Casing

A Formal Entryway — 80
Assembling the Door Surround
Building the Roof Unit
Installing the Entryway

Drilling a hole for a lockset →

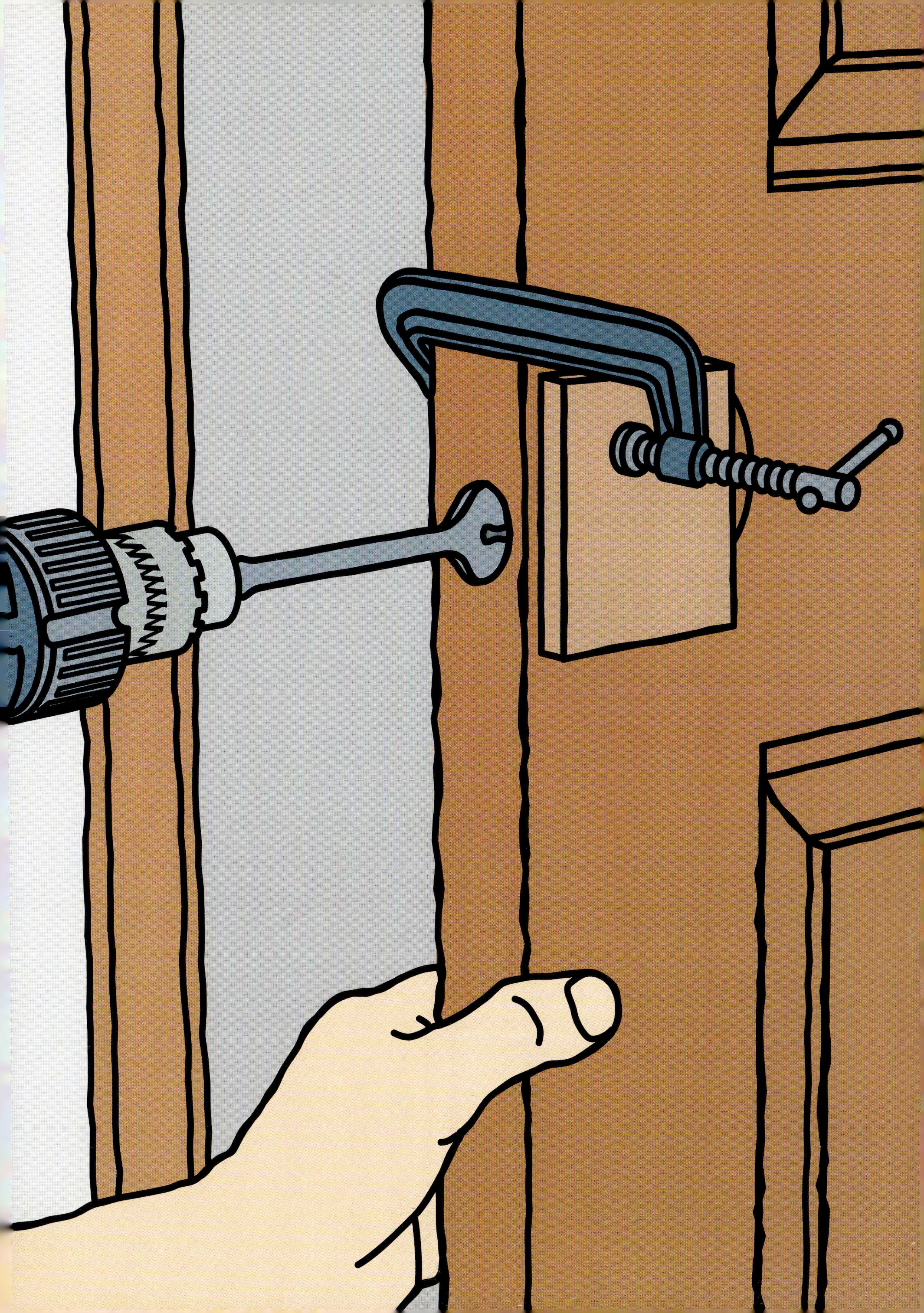

Building a Frame-and-Panel Door

Ready-made interior doors come in many styles, but usually only a few sizes. If you are replacing a standard-size door, you can probably find one for the job. But when you need to fit a door into an odd-size opening, or want it to match the other doors in the house, you can build one to suit. The traditional frame-and-panel style on these pages is attractive, sturdy, and relatively easy to construct.

Sizing the Door: Make the door $2\frac{3}{4}$ inches narrower and $1\frac{1}{2}$ inches shorter than the opening to allow for $\frac{3}{4}$-inch jambs, a $\frac{1}{2}$-inch space for shims on each side between the jambs and the rough framing, and $\frac{1}{8}$ inch clearance between the door and the jambs and threshold.

TOOLS
Table saw
Dado head
Featherboards
Clamps
Tenoning jig
Handsaw
Router
Router table
Panel-raising bit
Rubber mallet
Bar clamps
Paint scraper

MATERIALS
Lumber ($\frac{3}{4}$", $1\frac{3}{8}$")
Plywood ($\frac{1}{4}$")
Sandpaper (fine grade)
Wood glue

SAFETY TIPS
Protect your eyes with goggles when you are using a power saw.

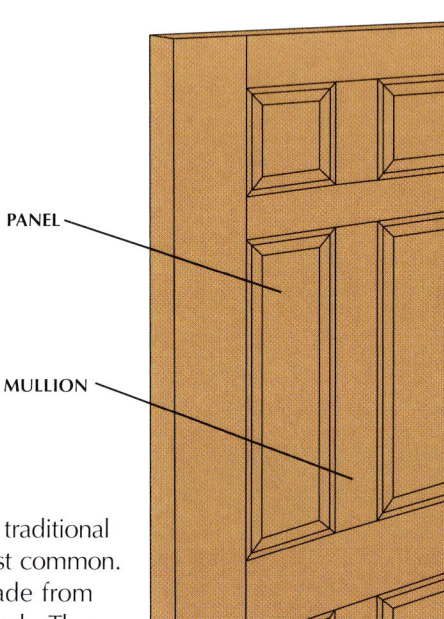

A six-panel door.
A door can have any number of panels, but the traditional six-panel design shown at right is one of the most common. The frames—stiles, rails, and mullions—are made from $1\frac{3}{8}$-inch lumber, and the panels from $\frac{3}{4}$-inch stock. The panels, which are "raised" on a router table so that they taper to a thin edge all the way around, fit in $\frac{1}{4}$-inch-deep grooves cut inside the frames, and the frame components are joined with plywood splines *(page 62)*. For extra strength at the door's four corner joints, larger splines are set in $\frac{3}{4}$-inch-deep grooves.

PANEL
MULLION
STILE
RAIL

MAKING GROOVES WITH A DADO HEAD

For cutting grooves in the door stock, the best tools are a table saw and an accessory called a dado head *(photograph)*. This device consists of two saw blades, sometimes sandwiched around smaller cutters, called chippers, and a number of washers. The two blades alone produce a $\frac{1}{4}$-inch cut. The chippers come in two thicknesses—$\frac{1}{8}$ and $\frac{1}{16}$ inch. Adding them between the blades will increase the cutting width of the dado head in $\frac{1}{8}$- or $\frac{1}{16}$-inch increments. You can add washers between the chippers or blades for finer width adjustment.

GROOVING THE FRAME PARTS

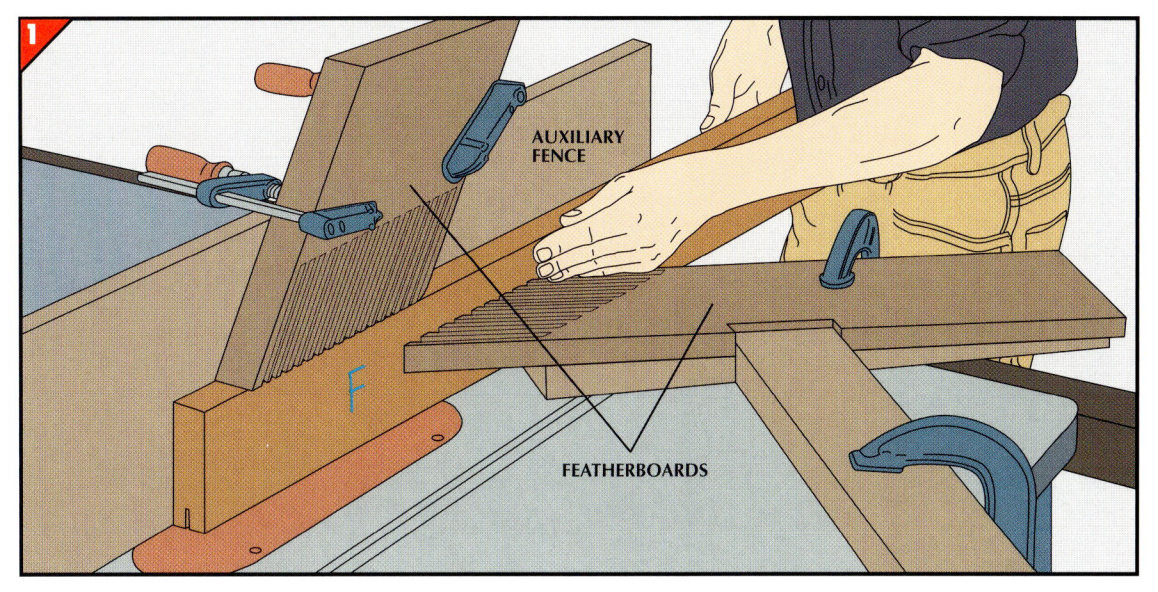

1. Preparing the stiles.
- On a table saw, cut the stiles to length and width, and label one face of each piece with an F for "front."
- Install a dado head (opposite, box) in the saw, adjusted to make a $\frac{1}{4}$-inch-wide kerf. Raise the blades to $\frac{1}{4}$ inch above the table.
- Attach a tall auxiliary wood fence to the saw's rip fence.
- Set a stile on the saw, front face out, and adjust the fence to center the blades on the edge of the stile. Support it with featherboards (page 10), centering the upper one over the blades.
- Turn on the saw and feed the workpiece through (above) at a steady rate until the end of the board is 3 inches from the blades. Then, move to the back of the table and pull the board past the blades while pressing it against the fence.
- Groove the other stile in the same way, then turn off the saw.
- To make the deep corner grooves, mark the center of the dado head on a piece of masking tape stuck to the saw table at right angles to the blades. Mark the planned positions of the inside edges of the top and bottom rails on the face of the stile. Raise the blades to $\frac{3}{4}$ inch above the table and feed one end of the stile into the blades until the mark on the stile lines up with the one on the masking tape. Pull the board back from the blades.
- Deepen the groove at the other end of the stile, then groove the second stile.

2. Preparing the rails and mullions.
- Cut the rails and mullions to size.
- Make a $\frac{1}{4}$-inch-deep groove along one edge of the top and bottom rails in the same way as for the stiles, then groove the remaining rails and the mullions along both edges.
- Raise the dado blades to $\frac{3}{4}$ inch above the table, then install a tenoning jig on the table and clamp the top rail in it, adjusting the jig to align the edge groove with the blades.
- Slide the jig to the front of the table, then turn on the saw and feed the workpiece into the blade (right). Pull the jig back and turn off the saw.
- Groove the other end of the top rail and both ends of the bottom rail in the same way, then lower the blades to $\frac{1}{4}$ inch above the table and groove the ends of the remaining rails and the mullions.

3. Test-fitting the joints.

◆ Cut four $\frac{1}{4}$-inch plywood splines to fit the corner joints; make two pieces the same length as the end of the top rail and two the length of the bottom rail, and all four slightly less wide than the combined depth of the mating grooves.

◆ Set a corner spline in position *(right)* and test-fit it— if it is too tight, sand down the spline and try again. Test-fit the other corner splines in the same way.

◆ Cut and test-fit splines for all the other joints between stiles, rails, and mullions in the same way.

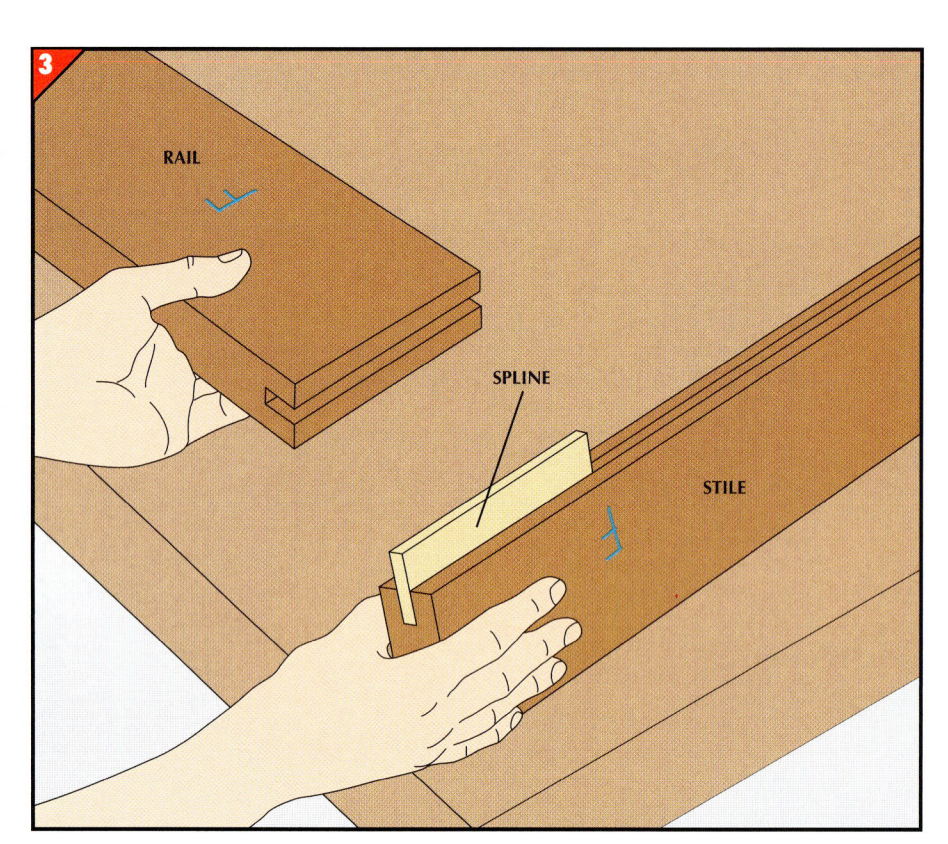

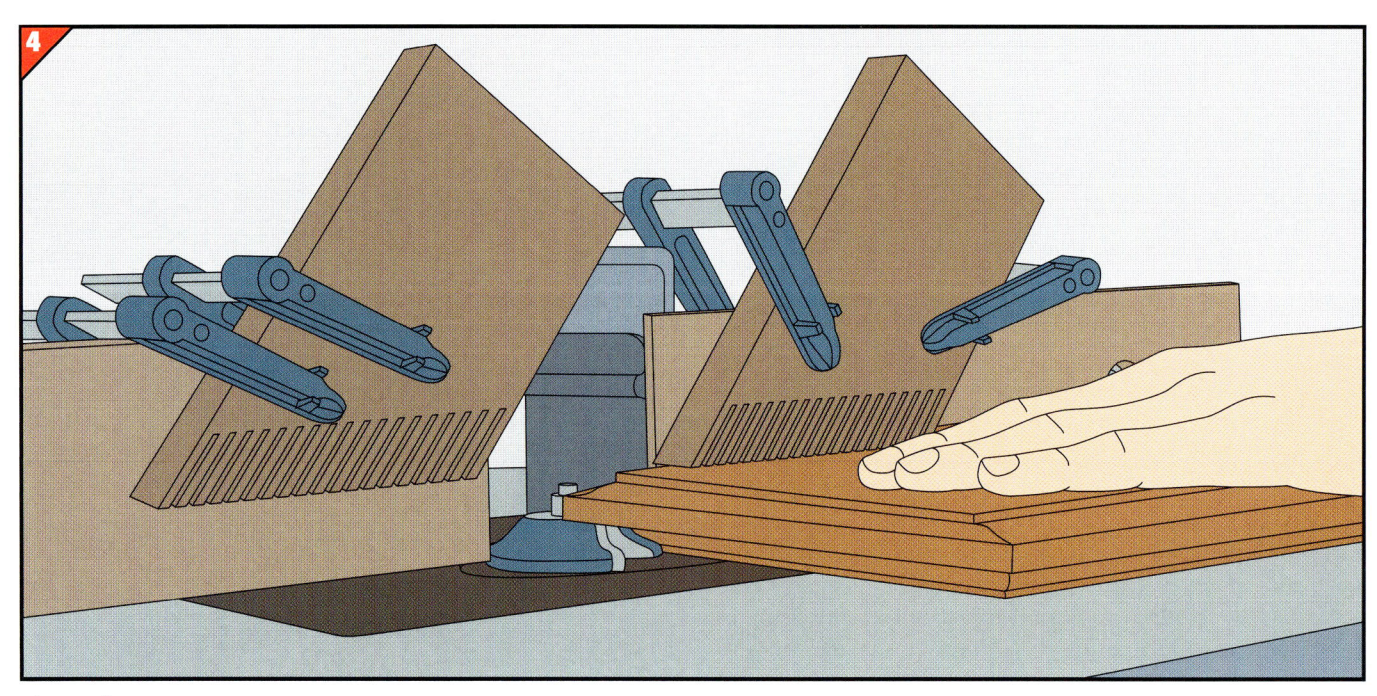

4. Raising the panels.

◆ From $\frac{3}{4}$-inch-thick lumber, cut each panel to the size of its frame plus $\frac{1}{4}$ inch all around to extend into the grooves.

◆ Install a piloted panel-raising bit in a router and mount it in a router table.

◆ Set the depth of cut at $\frac{1}{8}$ inch and clamp two featherboards to the fence, one on each side of the bit.

◆ To shape one end of the panel, feed the board across the table, keeping it flush against the fence and your hands clear of the cutter. Turn the panel over and repeat the cut on the same end, but on the opposite face.

◆ Shape the other end and both sides of the panel in the same way *(above)*.

◆ Raise the remaining panels, shaping their ends first and then the edges.

◆ Set the router bit to a depth slightly less than $\frac{1}{4}$ inch and make another pass all the way around both faces of the first panel, again starting with the ends and then shaping the edges.

5. Test-fitting a panel.
◆ Set the first panel into one of the grooves in a frame piece *(left)*. If the fit is too tight, raise the router bit slightly and make another pass all the way around the panel. Continue to test-fit the panel and shave wood from its perimeter until it fits.
◆ With the router bit at this height, make a second pass on all the remaining panels.
◆ Smooth the panels, stiles, rails, and mullions with fine sandpaper.

ASSEMBLING THE DOOR

1. Making a trial run.
◆ Without using glue or splines, assemble the door flat on a work surface in the sequence shown in the inset. Number the pieces as you go to aid in reassembly; on each stile and rail, mark the edges of the adjoining rail or mullion, then disassemble the door.
◆ Setting one of the stiles edge-up on the floor, apply wood glue to the portion of the groove that will house a rail—avoid getting any glue in the parts of the groove that will hold the panels. Also squirt glue into the groove in the end of the rail.
◆ Apply glue to the mating surfaces of the stile and rail. Insert splines at the rail position and fit the rail into the stile, aligned with the edge marks. Tap the ends of the rail lightly with a rubber mallet to close the joint.
◆ Reassemble the door pieces in this way, following the sequence and gluing in splines as you go *(right)*.

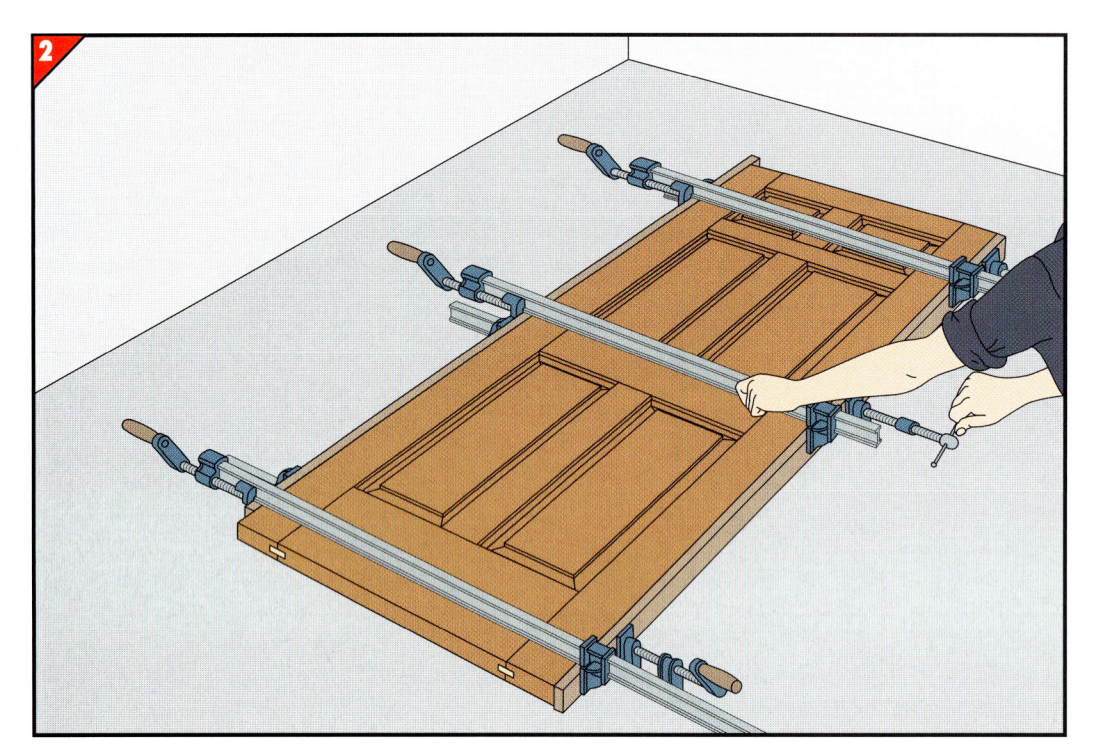

2. Clamping the door.

◆ Place three bar clamps on the floor, then carefully lay the assembled door on the clamps; align the clamps with the rails.
◆ With protective wood strips along the sides of the door beneath the jaws of the clamps, tighten the clamps just enough to close the joints uniformly.
◆ Install three more clamps across the top face of the door, also aligning them with the rails. Tighten all the joints until glue squeezes out of them, making sure there is the same pressure on the top and bottom clamps and that the door is flat *(above)*.
◆ Measure diagonally from corner to corner to check that the door is square and adjust the clamping pressure, if necessary.

Once the glue has dried, take off the clamps and remove any remaining adhesive with a paint scraper. When the glue has cured completely, sand and finish the door.

EMBELLISHING WITH MOLDING

The joinery method shown on the previous pages leaves a square shoulder where the panels enter the frame pieces. If you want a gentler profile, you can add quarter-round molding around the inside edges of the frame pieces *(right)*. Cut the molding to length, mitering it in the corners, and fasten it to the frame pieces—not the panels—with glue and 1-inch finishing nails.

Installing Jambs

Doors are held in their rough openings by a frame called a jamb. If you plan to install a ready-made door, choose a prehung unit, which is supplied with a factory-built jamb. When you build a door to fit an odd-size opening, you will have to make the jamb yourself.

Materials: For interior doors, purchase $\frac{3}{4}$-inch jamb stock; use 1- or $1\frac{1}{2}$-inch wood for exterior doors. In both cases, the stock should be as wide as the thickness of the wall. Special stock with grooves along the back face will prevent the wood from warping badly with humidity changes.

For the stops *(pages 67-68)*, choose flat stock since you can simply miter it to fit at the corners. With shaped stock, you will have to cope the ends *(page 15)*.

Installing the Jambs: A complete jamb assembly consists of a top and two side pieces joined with special joints called dadoes *(below)*. Here, grooves are cut into the side pieces to accommodate the top piece. You can make these grooves with a router whose bit matches the width of the stock, or you can use a table saw fitted with a dado head *(page 61)*. Plan the interior dimensions of the jamb $\frac{1}{4}$ inch wider and higher than the door itself.

For the door to hang properly, the jamb must be perfectly square and plumb—this is achieved by wedging shims between the jamb and the rough opening and making adjustments to them as necessary. If the finish floor is not yet in place, you will also need spacers under the side jambs to hold them until the unit is nailed in place.

TOOLS
Circular saw
Router
Electric drill
Screwdriver
Hammer
Carpenter's level
Plumb bob
Utility knife
Nail set
Combination square
Power or manual miter saw

MATERIALS
Jamb stock
1 x 4
Door stop
Shims
Straight board
Wood screws ($1\frac{1}{2}$" No. 8)
Finishing nails ($1\frac{1}{2}$", $3\frac{1}{2}$")
Wood glue
Spackling compound or wood putty

SAFETY TIPS

Put on goggles when you are driving nails or using a power tool.

MAKING AND FASTENING THE FRAME

1. Assembling the pieces.

◆ Cut the side jambs slightly shorter than the distance between the finish floor and the top of the rough opening. Make the head jamb $\frac{1}{2}$ inch longer than the desired inside width of the jambs.
◆ Measure along one side jamb a distance equal to the desired height of the jamb opening and make a mark. With a router or table saw, cut a dado $\frac{3}{4}$ inch wide and $\frac{1}{4}$ inch deep across the inside face of the jamb at and above this point. Mark and cut a dado on the other side jamb in the same way.
◆ Fit the head jamb into the dado in one of the side jambs and drill pilot holes for two $1\frac{1}{2}$-inch No. 8 wood screws. Fasten the jambs together with wood glue and screws.
◆ Fasten the head jamb to the other side jamb in the same way *(right)*.

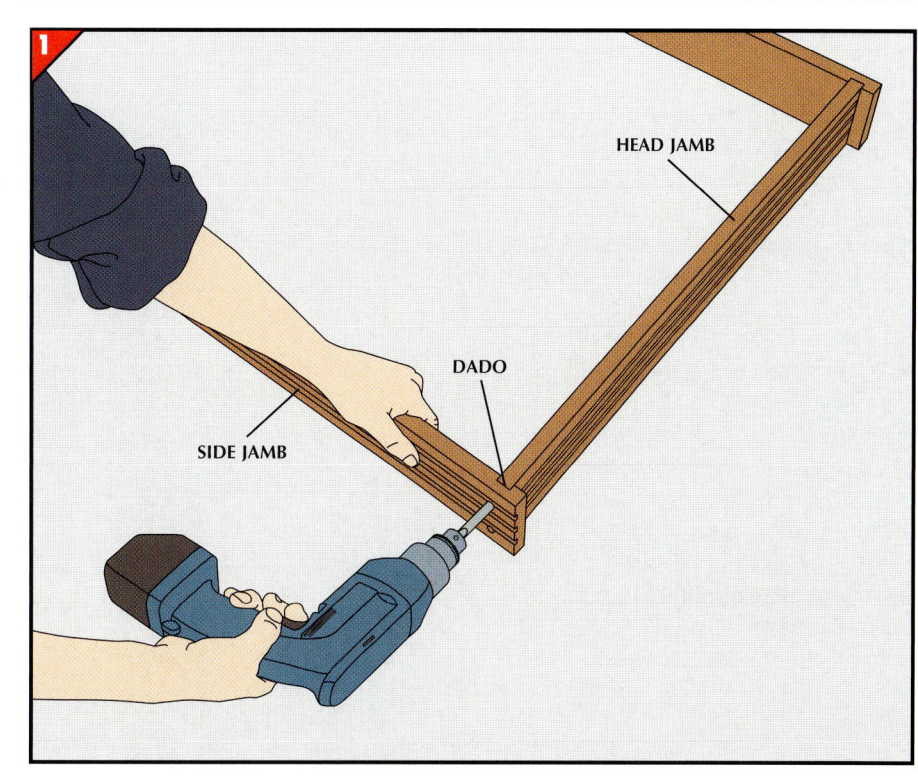

2. Positioning the jamb.

◆ Tack a brace to the wall diagonally across each top corner of the door opening.
◆ Position the jamb in the opening, propping it against the braces. Cut a 1-by-4 spreader to fit between the side jambs and place it on the floor between the jambs to keep them apart.
◆ Insert pairs of shims on both sides of the door between the side jambs and the rough framing at both ends of the head jamb, adjusting them to center the jamb assembly in the opening.
◆ With a carpenter's level, check the head jamb for level *(above)*; if necessary, shift the assembly slightly and adjust the shims.
◆ At the top of each side jamb, drive a $3\frac{1}{2}$-inch finishing nail through the jambs and shims into the rough framing.

3. Checking for plumb.

◆ Tap pairs of shims between the side jambs and the wall at both ends of the spreader.
◆ Mark the center of the head jamb on its edge and the center of the spreader on its face.
◆ Tack a nail into the edge of the head jamb at the center mark. Hang a plumb bob from the nail so the point of the bob is just above the spreader.
◆ Tap the shims flanking the spreader in or out to align the center mark on the spreader directly under the bob *(right)*.
◆ At the bottom of each side jamb, drive a nail through the jamb and shims into the rough framing, then remove the plumb bob, nail, and spreader.

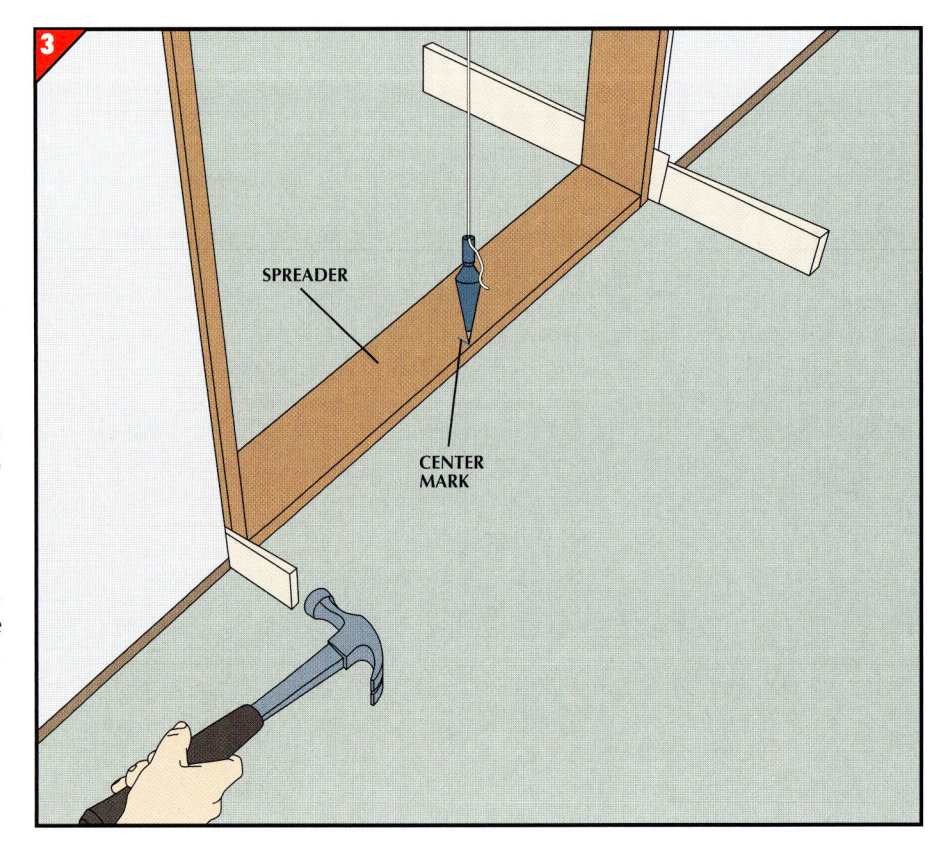

4. Squaring the side jambs.

◆ Drive additional pairs of shims at the hinge side of the jamb, locating one pair at each planned hinge location *(page 70, Step 1)*.
◆ Drive two pairs of shims on the lockset side, locating them just above and below the latch location *(page 76, Step 6)*.
◆ To ensure that the side jambs are straight, press a straight board or a long carpenter's level against one jamb to flatten it and nail through the jamb and shims into the rough framing *(right)*. Secure the opposite side jamb in the same way.
◆ Cut off the shims one at a time by holding the end of the shim and slicing across it repeatedly with a utility knife *(inset)* until you can break off the waste piece easily.
◆ With a nail set, sink all the nails; at each hinge location, bury the nails deeper than the thickness of the hinge leaves. If you will be painting the jambs, fill the holes with spackling compound; otherwise, apply wood putty.

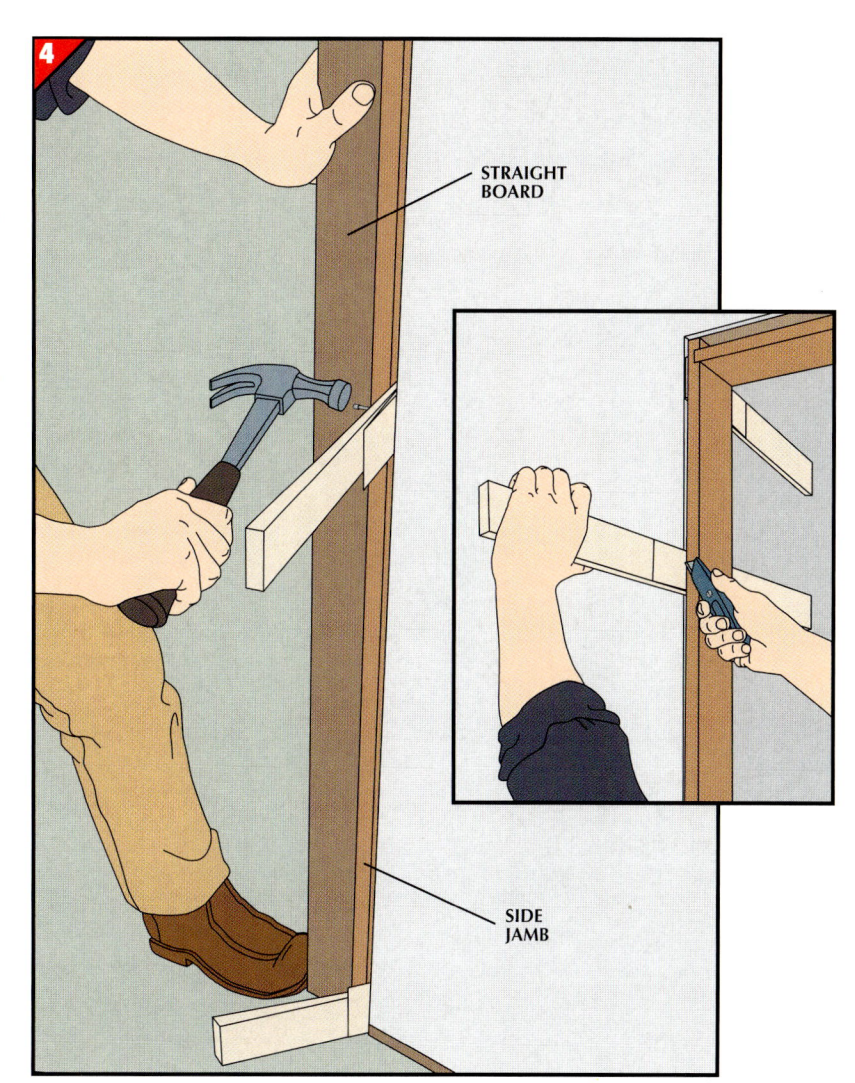

ADDING STOP MOLDING

1. Laying out the door stops.

◆ Adjust a combination square to the thickness of the door.
◆ On the side of the jamb that the door will sit flush with when closed, place the handle of the square against the edge of the lockset-side jamb. Set the tip of a pencil against the end of the ruler and run the square down the length of the jamb to mark a guideline for the door stop *(left)*.
◆ Mark a guideline on the hinge-side jamb in the same way, but add $\frac{1}{16}$ inch to the measurement to prevent the door from binding when it is closed.

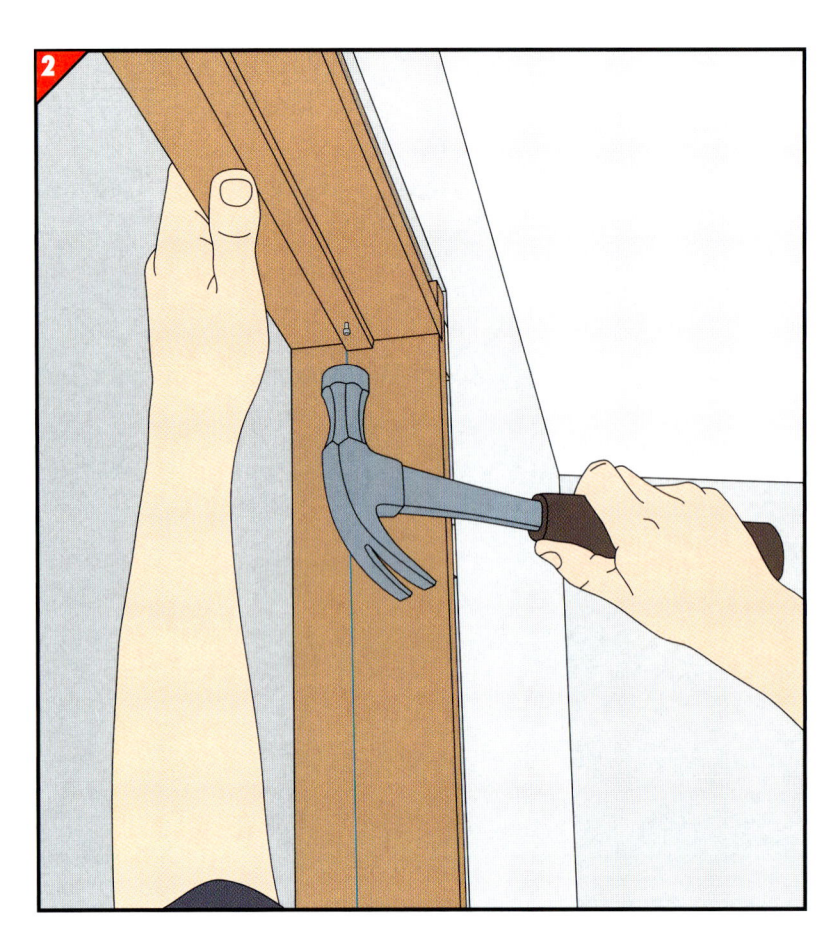

2. Installing the head-jamb door stop.

◆ Cut a length of door stop to fit along the head jamb, mitering both ends at 45 degrees with a power or manual miter saw.
◆ Position the stop on the head jamb, aligning the front edge with the guidelines on the side jambs. Tack the stop to the head jamb with 1½-inch finishing nails spaced every 10 to 12 inches and driven only partway in *(left)*.

3. Fitting in the side-jamb stops.

◆ Cut two lengths of door stop to fit along the side jambs, mitering the top ends.
◆ Align one stop with the guideline on the lockset-side jamb, holding the end tightly against the head stop. Tack it in place as you did the head-jamb stop *(right)*. Do not drive the nails home or install the hinge-side stop until you have hung the door *(page 72, Step 6)*.

Hanging the Door

Once the jambs are in place, you can install the door. The process involves fitting it to the opening and attaching the hinges.

Fitting the Door: If a ready-made door is too large, you can plane it to fit. A solid door can be trimmed by any amount, but do not remove more than 1 inch from the bottom or sides of a hollow-core type. If you are hanging a shop-made door *(page 60)*, you will need to cut a bevel along the lockset edge *(page 72)*.

Hinges: A solid-core door, or any type taller than 80 inches, requires three hinges. Hollow-core doors or solid ones shorter than the standard can be hung with only two. The size of the hinges depends on the width and thickness of the door *(chart, below)*.

To attach the hinges, you first need to cut shallow insets called mortises in the door and jamb. The best tool for making them is a router *(page 70, Step 1)*. On both the jamb and door, cut the mortises as long as the hinges and $\frac{1}{4}$ inch narrower than the width of the hinge leaves. When installed, the leaves will extend past the contacting edges of the door and jamb by $\frac{1}{4}$ inch, preventing the hinges from binding. If you choose hinges with square rather than rounded corners, you will need to chisel out the corners to fit the hinges.

TOOLS
Saber saw
Electric drill
Countersink bit
Screwdriver
C-clamps
Router
Template guide
Wood chisel
Utility knife
Awl
Jack plane
Nail set
Hammer
Block plane

MATERIALS
1 x 2s
2 x 6 scraps
Plywood ($\frac{3}{4}$")
Wood blocks
Wood screws (2" No. 6)
Common nails ($2\frac{1}{2}$")
Hinges
Spackling compound or wood putty

SAFETY TIPS
Wear goggles when using a power tool.

Door Thickness	Door Width	Hinge Height
$1\frac{3}{8}$"	Up to 32"	$3\frac{1}{2}$"-4"
	More than 32"	4"-$4\frac{1}{2}$"
$1\frac{3}{4}$"	32"-36"	5"
	36"-48"	5" (heavy-duty type)
	More than 48"	6"

Choosing hinges.
Determine the required height of the door hinges according to the width and thickness of the door, as specified in the chart at left. The width of the hinges varies with the door's thickness; for a door up to $1\frac{3}{8}$ inches thick, you need hinges 3 inches wide; for a $1\frac{3}{4}$-inch-thick door, $3\frac{1}{2}$-inch-wide hinges are required.

ATTACHING HINGES

1. Routing the jamb mortises.

◆ Purchase a hinge-mortising jig *(opposite)*, or make a template to guide the router: With a saber saw, cut a piece of $\frac{3}{4}$-inch plywood about 6 by 12 inches. Make a cutout centered along one edge, cutting it as long as a hinge leaf; for its width, make it as wide as the planned mortise, adding the diameter of the router bit's template guide and $\frac{3}{4}$ inch for the thickness of the fence. Cut a 2-inch-wide fence the same length as the template and fasten it to the template with four countersunk 2-inch No. 6 wood screws *(right, top)*.

◆ Mark the top of the upper hinge on the side jamb by measuring down 7 inches from the head jamb. Mark the bottom of the lowest hinge 11 inches above the bottom of the jamb. If you are adding a third hinge, locate it halfway between the other two.

◆ Clamp the template to the jamb so the fence is against the edge of the jamb on the side where the door will open, and the top of the cutout is aligned with the upper-hinge mark.

◆ Fit the router with a straight bit and template guide, then adjust the depth of cut to the combined thickness of the template and the hinge leaf. With the router flat on the template, move it in small clockwise circles within the template cutout *(right, bottom)* until the bottom of the mortise is flat.

◆ Reposition the template and rout the middle and bottom mortises.

◆ For rectangular hinges, square the corners of the mortises with a wood chisel.

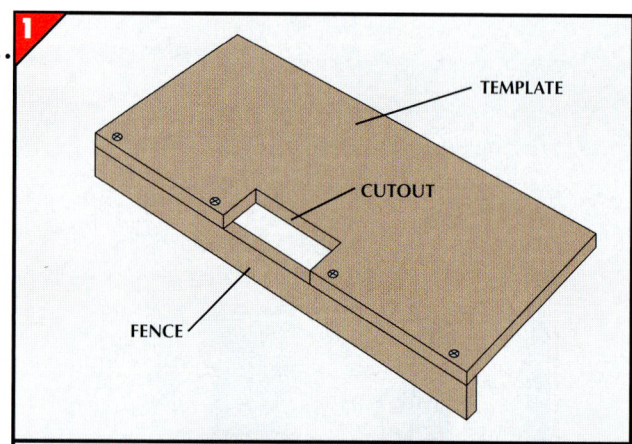

2. Transferring the hinge locations.

◆ Working with a helper, prop the door in its frame.

◆ With your helper holding $2\frac{1}{2}$-inch nails as spacers between the top of the door and the head jamb, drive wood blocks under the door to wedge it tight against the nails.

◆ Drive a shim between the door and the lockset-side jamb about 3 feet from the floor to push the door against the hinge-side jamb.

◆ With a utility knife, nick the edge of the door at the top and bottom of each hinge mortise in the jamb *(left)*.

◆ With the router and template, rout the hinge mortises on the door, aligning the cutout in the template with the knife cuts.

A HINGE-MORTISING JIG

A commercial hinge-mortising jig has templates to fit three different hinge sizes and spacings. Adaptable to doors either $1\frac{3}{8}$ or $1\frac{3}{4}$ inches thick, the jig can be used for mortises on both the jamb and the door. In the model shown here, plastic spacers are inserted in the jig to change the size of the hinge cavity.

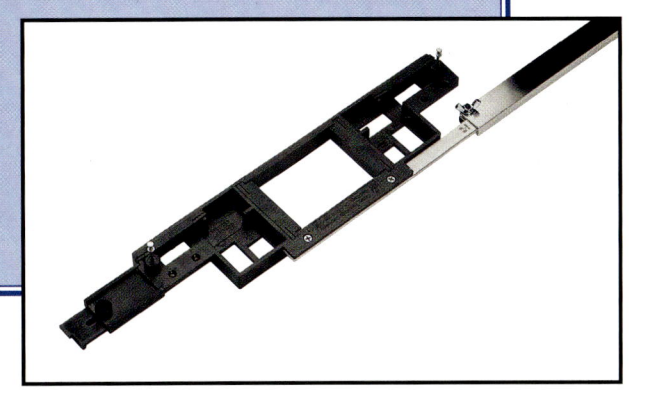

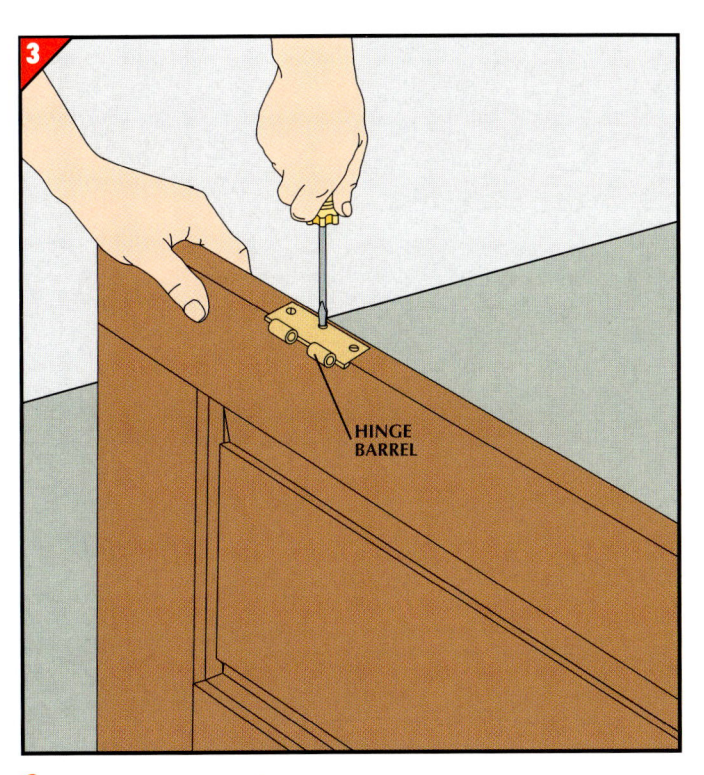

3. Attaching the hinges.
◆ Separate the two leaves of a hinge by pulling out the pin. Set the leaf with two barrels in a mortise on the door and mark the screw holes with an awl.
◆ Remove the hinge leaf, drill pilot holes for the screws provided, and fasten the leaf to the door *(above)*.
◆ Install the other hinges on the door in the same way, then attach the matching leaves to the jamb.

4. Hanging the door temporarily.
◆ Lift the door into position and slip some wood blocks under it. Shift the door to engage the barrels of the top hinge, then slide the hinge pin in partway *(above)*.
◆ Pivot the door to join the bottom hinge leaves and slip the hinge pin in partway, then fit the pin in the middle hinge.

5. Marking the bevel.
◆ Standing on the door-stop side, close the door—its front edge will hit the edge of the jamb, preventing it from closing fully.
◆ Holding the door against the jamb, scribe a pencil line down the face of the door where it meets the jamb *(right)*.

6. Beveling the door edge.
◆ With a ruler, extend the pencil line across the ends of the door, marking the angle of the bevel.
◆ Holding a jack plane at the same angle, guide it along the door edge *(left)*. A portable power planer that can be set to the desired bevel angle—3 to 5 degrees—is also handy for this job *(photograph)*.
◆ Continue planing until you reach the pencil line.
◆ Rehang the door as in Step 4, inserting the hinge pins all the way this time.
◆ Install the hinge-side door stop, butting it against the head-jamb stop *(page 68, Step 3)*.
◆ Pass a dime along the hinge jamb to check for the correct clearance between it and the door; use a nickel to check the clearance at the top and other side of the door. If necessary, take the door down again and plane any high spots. Use a block plane for the top or bottom of the door, working from the edges toward the center to avoid splintering the end grain.
◆ When the door fits properly, drive all the nails home and sink them with a nail set, then fill the holes with spackling compound for jambs to be painted, or wood putty if you will be staining the door.

Installing a Lockset

Putting in a lockset is one of the easiest parts of installing a door; however, it does require precise measuring to ensure that the door latches properly.

Modern locksets are categorized as tubular or cylindrical. The mechanisms of the two differ somewhat—the tubular type shown on these pages is the simplest—but their installation is virtually the same.

Positioning the Unit: Locksets are typically set in the door with the center of the knob 36 inches from the floor and, depending on the model, either $2\frac{3}{8}$ or $2\frac{3}{4}$ inches from the latch edge of the door. A template is usually provided with the hardware to help with accurate placement of the holes.

TOOLS

Awl
Electric drill
Hole saw
Spade bit
C-clamp
Wood chisel
Screwdriver
Combination square

SAFETY TIPS

Protect your eyes with goggles when using an electric drill.

PUTTING IN A PRIVACY LOCK

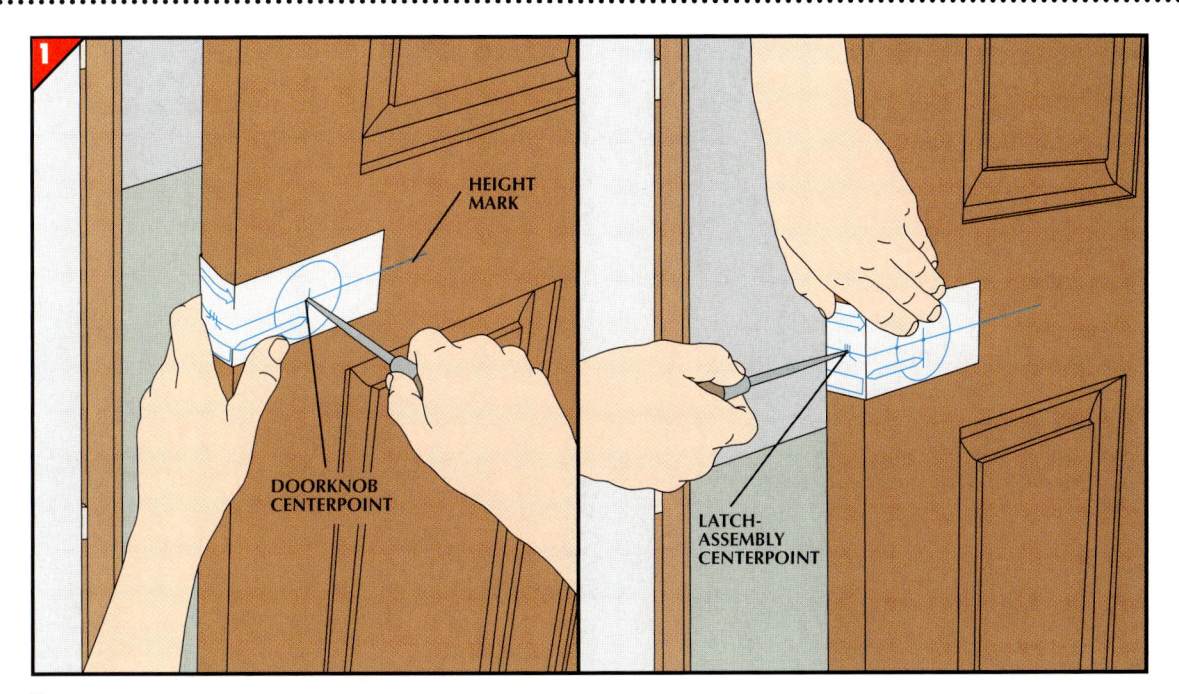

1. Positioning the knob assembly.
◆ Measure up 36 inches from the floor and mark a horizontal line on one face of the door.
◆ Tape the template provided with the lockset to the door, aligning the doorknob centerpoint on the template with the height mark.
◆ With an awl, mark the centerpoint on the face of the door *(above, left)*.
◆ Mark the centerpoint for the latch-assembly hole on the door's edge *(above, right)*.

2. Drilling the knob hole.

◆ In an electric drill, install a pilot-bit-type hole saw of the diameter specified on the template.
◆ Place the bit's point on the awl mark you made on the door's face in Step 1, then drill into the door *(left)* until the point of the pilot bit emerges from the other side.
◆ Working on the other side of the door, insert the bit in the hole and keep drilling until the hole saw has cut completely through the door.

3. Boring the latch-assembly hole.

◆ To prevent the door from splitting, clamp a wood block to each side of the door over the doorknob hole and flush with the latch edge.
◆ Fit the drill with a spade bit of the diameter indicated on the template, set the tip of the bit on the awl mark on the door edge, and drill the latch-assembly hole *(right)*. Depending on the lockset model, stop drilling when the bit reaches the doorknob hole, or continue beyond it if required for clearance.

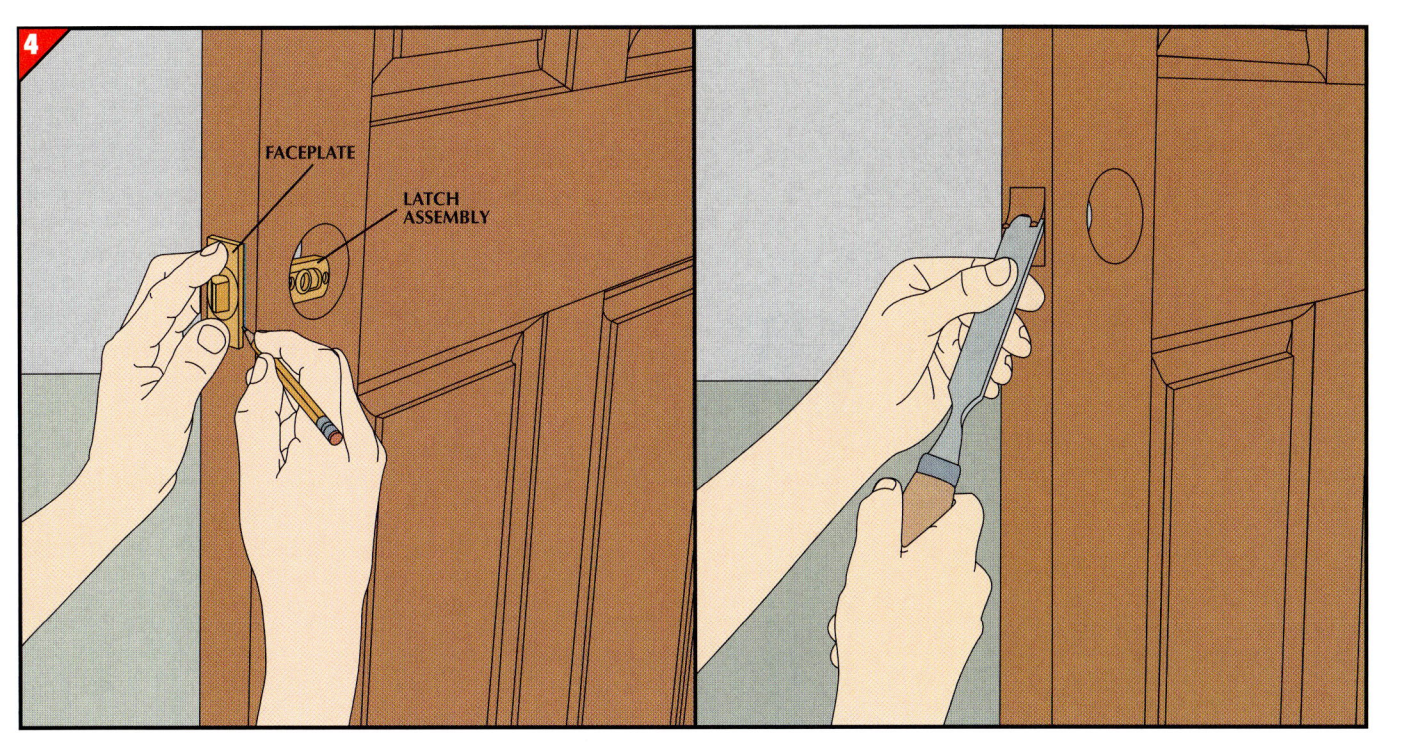

4. Attaching the latch assembly.

◆ Slide the latch assembly into its hole in the door so the beveled side of the latch bolt will hit the jamb first as the door closes, and the faceplate is against the door edge.
◆ Hold the faceplate square to the edge and trace its outline with a pencil *(above, left)*. Remove the latch assembly.
◆ Score the outline with a chisel. Then, holding the blade bevel-side down, pare out the waste *(above, right)*, creating a mortise to fit the faceplate. Test-fit the plate in the mortise periodically, stopping when the plate is flush with the door edge.
◆ Hold the faceplate in place and mark the screw holes with an awl. Remove the assembly, drill pilot holes for the screws provided, and fasten the assembly to the door.

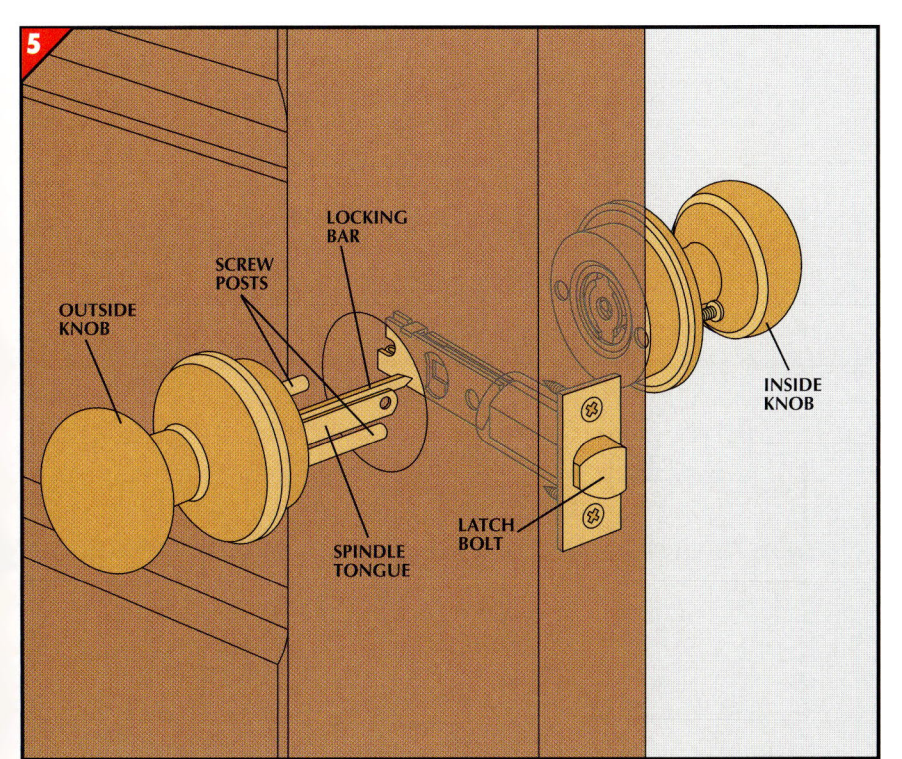

5. Installing the doorknobs.

◆ Depress the latch bolt about $\frac{1}{8}$ inch and slip the outside knob into its hole, sliding the screw posts, spindle tongue, and locking bar through their respective holes in the latch assembly.
◆ Slide the inside knob in place over the spindle tongue, locking bar, and screw posts, then fasten the inside knob to the outside one with the screws provided.

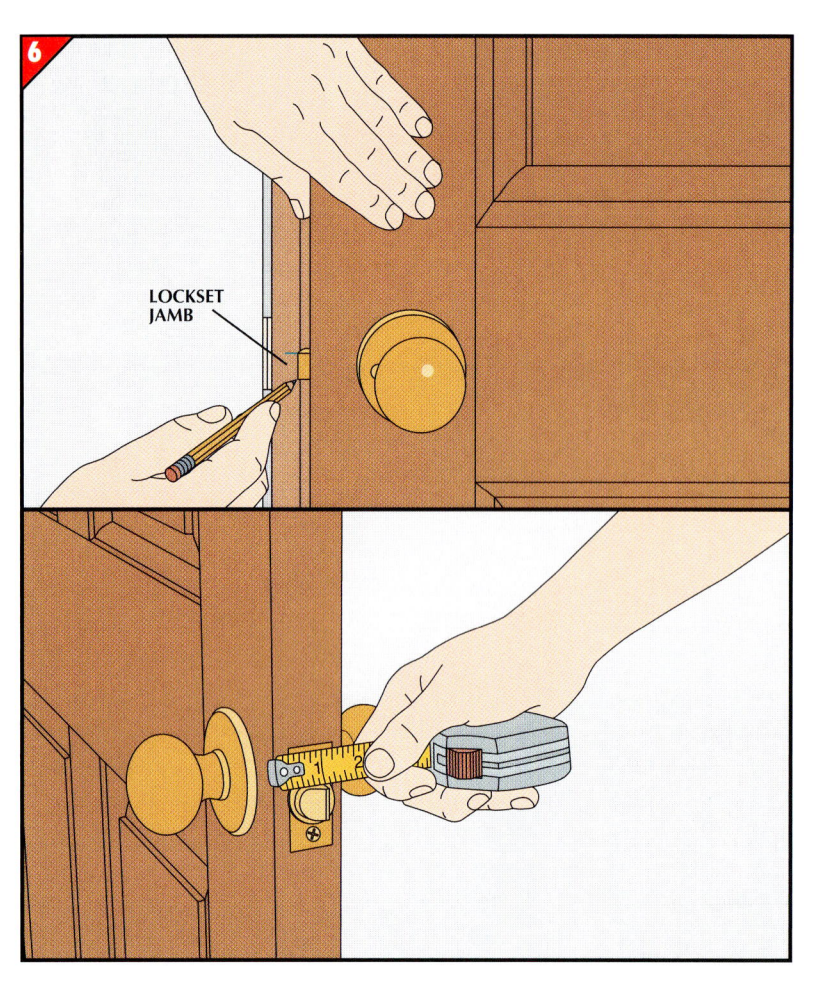

6. Positioning the latch hole.
◆ Partially close the door so the latch is touching the edge of the lockset jamb, then mark the top and bottom of the latch on the jamb *(left, top)*.
◆ With a combination square, extend the marks to the face of the jamb.
◆ Measure the distance from one face of the door to each edge of the latch bolt *(left, bottom)*. Transfer these measurements to the jamb face and mark vertical lines across the horizontal ones to outline the latch bolt.
◆ Score the outline with a chisel, then use the chisel or a drill with a spade bit to make a mortise in the jamb deep enough to accommodate the latch bolt.

7. Fastening the strike plate.
◆ Center the strike plate over the latch hole in the jamb so the tongue extends past the side of the jamb and the screw holes are outside the bolt mortise. Outline the plate on the jamb *(right)*.
◆ Chisel a mortise within the outline *(page 75, Step 4)* as deep as the strike-plate thickness.
◆ Hold the strike plate in position and mark the screw holes on the jamb. Drill pilot holes at the marks for the screws provided and fasten the plate to the jamb.
◆ Close the door. If it does not latch properly, you can adjust the fit by bending the strike-plate tongue slightly in or out.

Finishing Up with Trim

The final step in installing the door is to add casing. This molding hides the gap between the jamb and the wallboard, and gives the door a finished look.

Casing Styles: It's best to use the same style of molding for the door as for the windows in the room. If stool-and-apron trim was applied around the windows *(pages 53-57)*, make butt joints in the door casing *(below)*. For a room with windows trimmed in picture-frame style *(pages 46-52)*, miter the corners of the door molding following the same technique.

Preparing the Jamb: Since the casing links the jamb and the wall, these two surfaces must be flush. If the jamb is slightly proud of the wall, plane it down. Where it is slightly shy of the wall, shave down the wallboard with a rasp *(page 48, Step 1)*; but if the jamb is set back more than $\frac{1}{4}$ inch from the wall, make and install a jamb extension of the same type used for a stool-and-apron window *(page 55, Step 4)*.

TOOLS
Combination square
Hammer
Circular saw
Power or manual miter saw
Nail set

MATERIALS
Finishing nails ($1\frac{1}{2}$", 2")
Spackling compound or wood putty

SAFETY TIPS

Put on goggles when hammering or when using a power tool.

Butted door casing.

In this classic casing style, side and head casing meet with a simple butt joint, and the head casing overhangs the side casing by about $\frac{3}{4}$ inch at each end. The inside edges of the casing are offset from those of the jambs, leaving a $\frac{1}{4}$-inch reveal. Decorative rosettes can be added at the top corners if desired *(page 53)*.

Rather than resting on the floor, the side casing usually rests on a pair of plinth blocks slightly thicker and wider than the casing and 1 inch taller than the baseboard. If you choose to install flat plinth blocks like those shown here, they can be decorated with a slight bevel; or, you may want to buy blocks with a profile matching that of the casing.

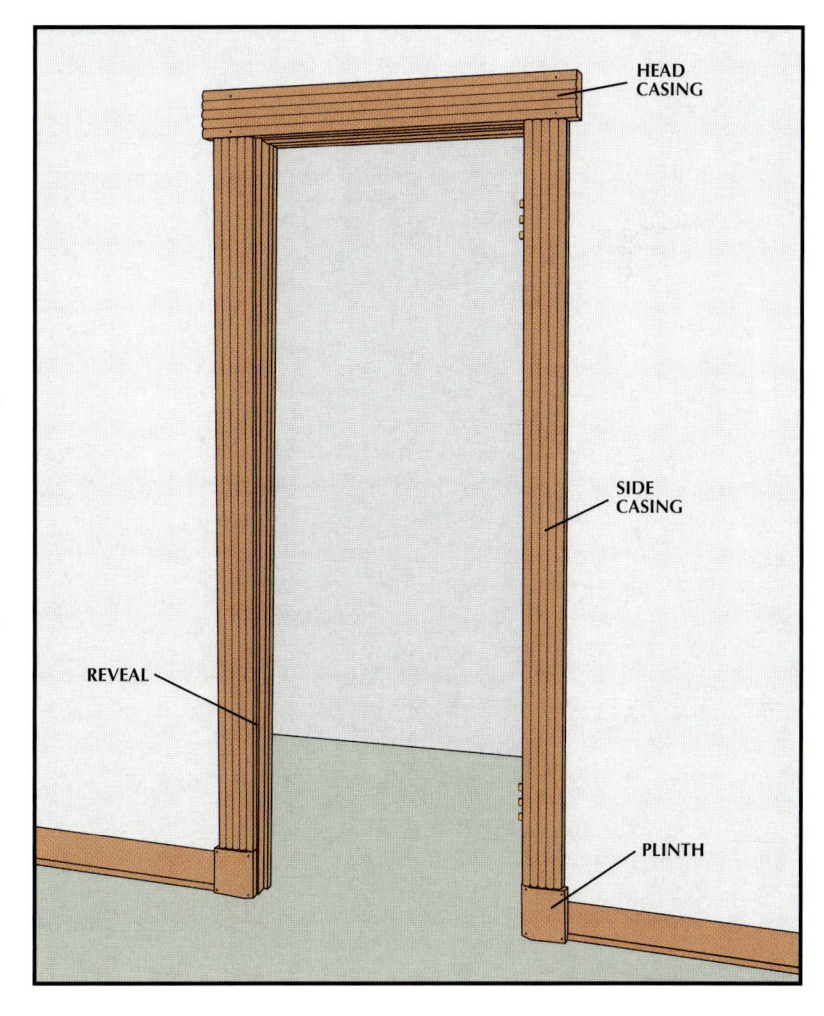

ATTACHING THE CASING

1. Marking the reveal.
◆ Remove the door.
◆ Set a combination square to $\frac{1}{4}$ inch, checking that this is wide enough to clear the hinges. Butting the handle of the square against the face of a side jamb and resting a pencil against the end of the ruler, mark a reveal line the length of the jamb edge *(right)*. A shop-made reveal gauge can also be used *(page 48)*.
◆ Mark the same reveal on the other side jamb as well as on the head jamb.

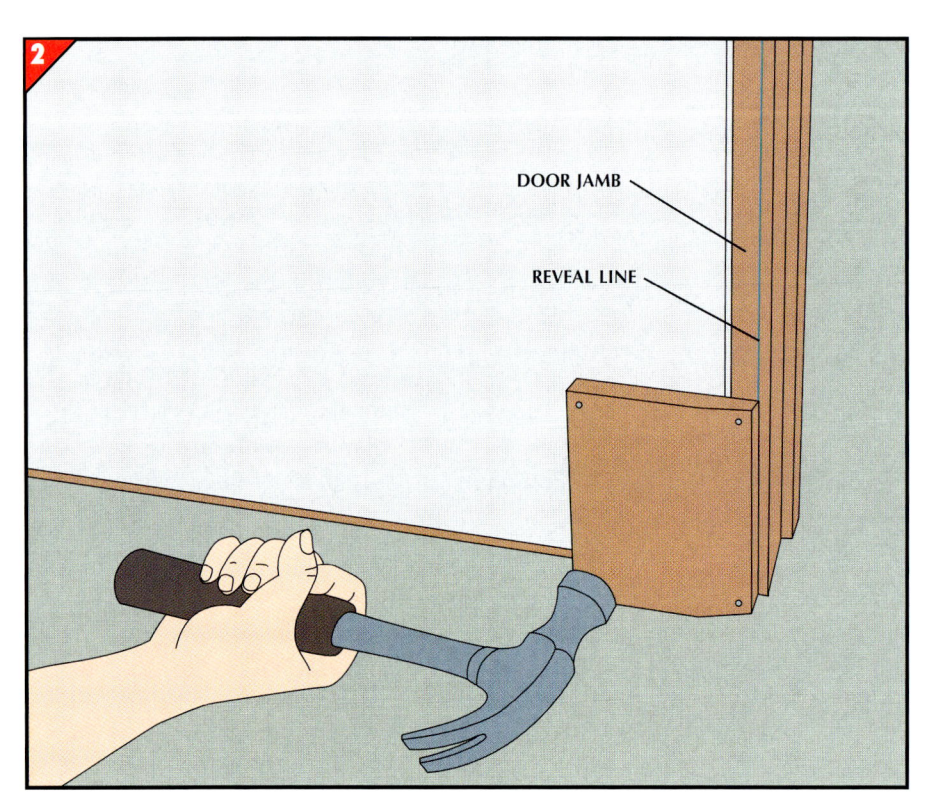

2. Fastening the plinths.
◆ Cut two plinth blocks to the desired height and width *(page 77)*. If desired, bevel one edge of the blocks with a power or manual miter saw; to cut larger stock with a power saw, you may need a compound miter saw.
◆ Align the edge of a block with the reveal line—if the finish flooring is not yet in place, set a scrap of wood the thickness of the flooring under the plinth block.
◆ For softwood lumber, nail the plinth block to the jamb with two $1\frac{1}{2}$-inch finishing nails and to the rough framing through the wallboard with two 2-inch nails *(left)*; for hardwood stock, drill pilot holes first.

3. Attaching the head casing.

◆ Measure the distance between the reveal lines on the two side jambs. Add to this figure the width of both of the side casings to be used plus $1\frac{1}{2}$ inches and cut the head casing to this length.
◆ Mark the center of the head jamb and the head casing. Align the head casing with the reveal mark and line up the two center marks.
◆ Nail the casing to the jamb with $1\frac{1}{2}$-inch finishing nails and to the rough framing through the wallboard with 2-inch nails, spacing the nails every 6 inches *(right)*—drill pilot holes first if you are installing hardwood casing.

4. Fitting in the side casing.

◆ Cut the side casings to fit snugly between the plinths and head casing.
◆ Fit one piece into position, aligning it with the reveal line *(left)*. Fasten it in the same way as the head casing, then put up the other side piece.
◆ With a nail set, sink all the nails. If you plan to paint the casing, fill the holes with spackling compound; otherwise use wood putty.

A Formal Entryway

The plain exterior of a house can be transformed with a custom-built entryway. The example shown below incorporates vertical pilasters and a head unit that appear to support a small roof. It has a base of plywood to which door casing and other wood moldings are fastened. The sections are assembled in the workshop, and then installed.

Choosing Materials: The base is made of exterior-grade plywood to better resist the elements. The exposed edges of the plywood are concealed with picture-frame trim. If only finger-jointed rather than solid-wood trim is available, substitute half-round molding. Use galvanized fasteners throughout and treat the entire unit with wood preservative before priming and painting it.

Adapting the Design: The simple design shown on these pages will blend with a variety of architectural styles. For a more elaborate look, you can replace the pilasters with ready-made half columns, or install a prefabricated semicircular sunburst instead of the peaked roof *(page 86)*.

The unit shown in this section is sized to accommodate a door opening measuring 35 inches wide and 81 inches high. If your opening is a different size, adjust the dimensions as required to maintain the proportions and the correct alignment of the various elements.

TOOLS

Table saw
Carpenter's square
Power or manual miter saw
Combination square
Electric drill
Doweling jig
Combination bit
Masonry bit
Circular saw and edge guide
Adjustable T-bevel
Clamps
Nail set
Pry bar
Caulking gun

MATERIALS

1 x 3, 1 x 4, 1 x 8
2 x 4
Exterior plywood ($\frac{3}{4}$")
Plywood ($\frac{3}{4}$")
Shims
Door casing ($3\frac{1}{8}$")
Door-stop molding ($\frac{1}{2}$")
Crown molding ($2\frac{1}{2}$")
Picture-frame molding ($\frac{3}{8}$")
Galvanized finishing nails (1", $1\frac{1}{4}$", $1\frac{1}{2}$", 2")
Galvanized wood screws ($1\frac{1}{2}$", 2" No. 8)
Masonry screws (2")
Exterior-grade wood glue
Spackling compound
Caulk

SAFETY TIPS

When hammering or using power tools, protect your eyes with goggles.

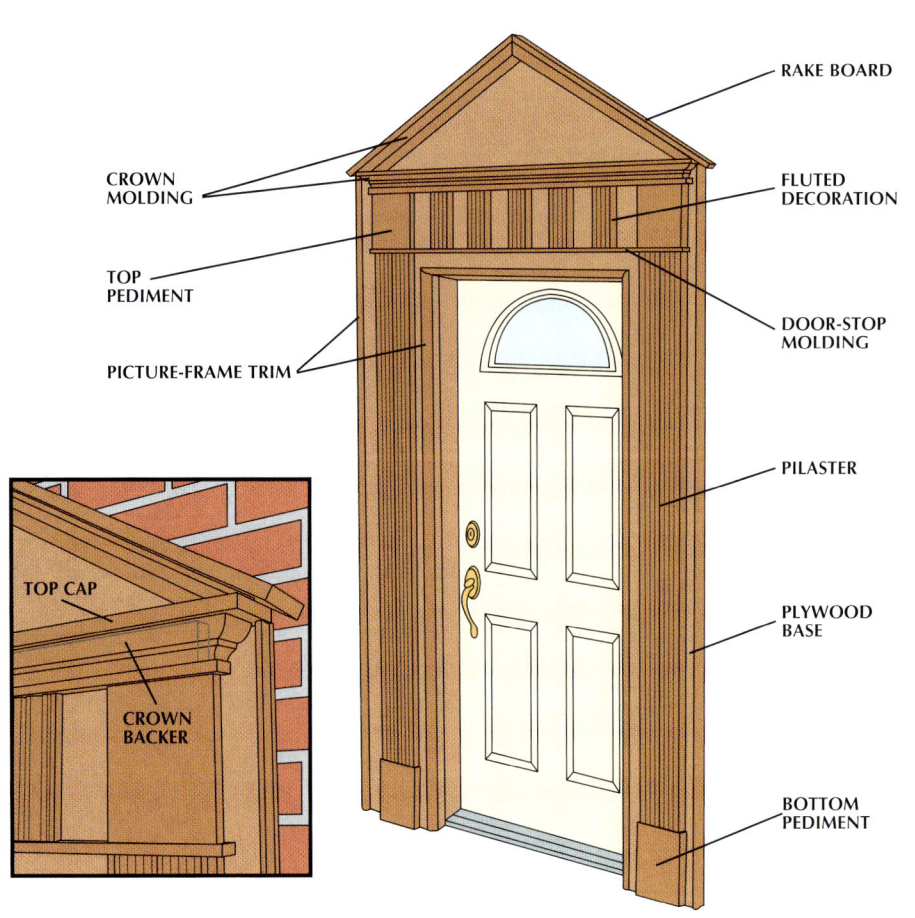

Anatomy of a formal entryway.

The base for the door surround is three plywood sections fastened together. Bottom pediments support fluted pilasters made of two lengths of symmetrical door casing fastened side by side. Resting on the pilasters is a head unit featuring fluted decorations of the same door casing, top pediments, and horizontal door-stop molding. A top cap is fastened to the top of the base, and crown molding resting against a backer fits between it and the roof unit. Picture-frame molding conceals the exposed plywood edges.

The roof unit consists of a triangular plywood base. Sloping rake boards are fastened to the plywood and trimmed with crown molding.

ASSEMBLING THE DOOR SURROUND

1. Joining the base.
◆ From $\frac{3}{4}$-inch exterior plywood, cut two side pieces $10\frac{1}{4}$ inches wide. To determine their length, measure the distance from the inside of the head jamb to the door stoop and add $\frac{1}{4}$ inch—the width of the picture-frame trim at the top less the width of the shims needed to install the surround. Cut the pieces to that length ($81\frac{1}{4}$ inches in the example shown).
◆ Cut a top piece $12\frac{1}{2}$ inches wide. For its length, measure the distance between the two side jambs and add the width of the two side pieces plus two pieces of picture-frame trim. Cut the top piece to that length (here, $54\frac{1}{4}$ inches).
◆ Create a work surface out of plywood that is large enough to lay out the base, with room for strips of plywood screwed to the top and bottom, leaving small gaps between the strips and the base pieces *(inset)*. If the base unit is too large for one sheet of plywood, splice another short piece to the first with cleats.
◆ Drill holes for dowels *(page 37)* to join the side pieces to the top. Lay wax paper under the joints, then apply exterior-grade wood glue to the dowels, holes, and matching edges, and lay the pieces in position on the work surface.
◆ Fill the gaps between the plywood end strips and the base with scrap wood, then drive shims between the plywood strips and the scrap wood to force the base pieces together *(right)*, alternating between the two sides until the pieces are snug. As you work, check the alignment occasionally with a carpenter's square.
◆ Once the glue is dry, remove the shims, plywood strips, and scrap wood. Fasten a plywood brace across the bottom of the unit to keep it square.

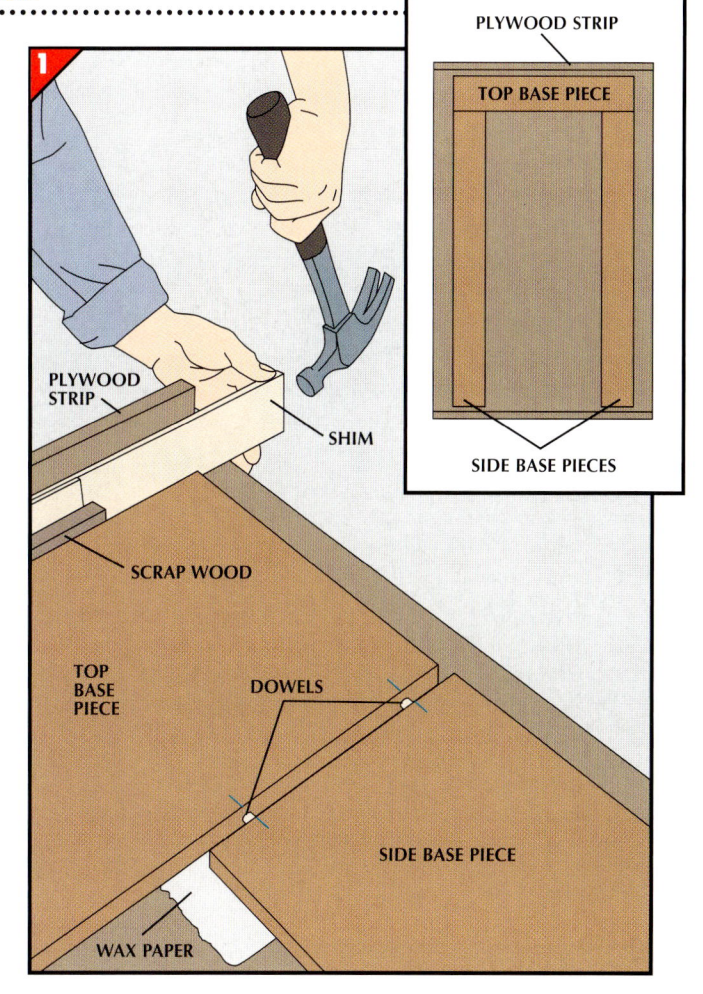

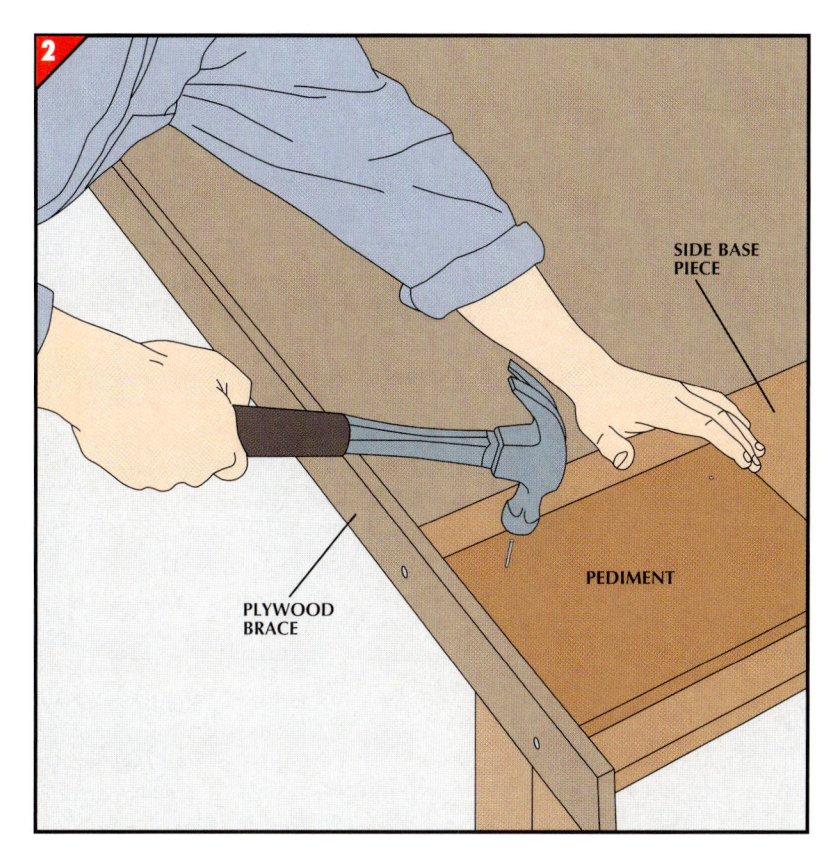

2. Attaching the bottom pediments.
◆ For the bottom pediments, cut two 10-inch-long sections from a 1-by-8, then trim them $\frac{3}{4}$ inch wider than two pieces of the door casing you will be using *(Step 3)*; for $3\frac{1}{8}$-inch-wide casing, the width would be 7 inches.
◆ Align one pediment with the bottom of a side base piece, and center it on the piece.
◆ Fasten the block with exterior-grade wood glue and four $1\frac{1}{4}$-inch galvanized finishing nails *(left)*.
◆ Attach the other bottom pediment in the same way.

3. Fastening the pilasters.

◆ Cut $3\frac{1}{8}$-inch-wide symmetrical door casing into four lengths—$73\frac{1}{4}$ inches, in this case—so each piece is 8 inches shorter than the length of the side base pieces and extends 2 inches onto the top base piece.
◆ With a combination square, mark a vertical line 2 inches from the edge of the side base piece *(right)*.
◆ Align a length of door casing with this line, butting it against the top of the bottom pediment. Fasten the casing to the base with glue and two rows of nails starting 2 inches from the end of the casing and spaced about every foot.
◆ Nail a second length of door casing alongside the first.
◆ Fasten casing to the other side base piece in the same way.

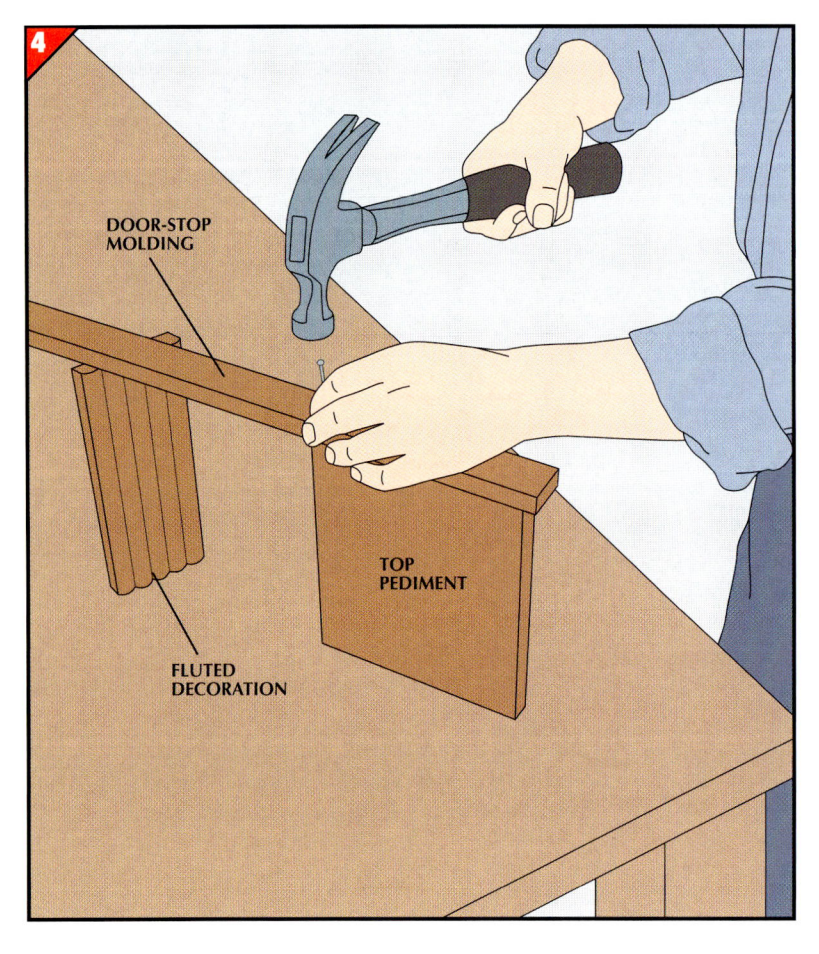

4. Assembling the head unit.

◆ On a table saw, rip two lengths of $\frac{1}{2}$-inch-thick door-stop molding $1\frac{1}{4}$ inches wide. Make the pieces—$53\frac{1}{4}$ inches in this example—3 inches shorter than the length of the top base piece.
◆ From a 1-by-8, cut two top pediments $7\frac{5}{8}$ inches long and as wide as the pilasters ($6\frac{1}{4}$ inches in this case).
◆ Cut five $7\frac{5}{8}$-inch-long pieces of the casing used for the pilasters to make the fluted decorations.
◆ Lay out the head assembly on a work surface, with a pediment $\frac{1}{2}$ inch from each end of the lengths of molding and the five decorative pieces spaced equally in between. Mark the location of the pediments and decorations on the molding.
◆ Set one of the pieces of molding on top of the two pediments, carefully aligning them so they are flush with one edge of the molding. Prop up the molding with the fluted decorations. Fasten the molding to the pediments with two $1\frac{1}{4}$-inch nails *(left)*.
◆ Slip the fluted decorations into position, also flush with one edge of the molding, and fasten the molding to them.
◆ Turn the assembly over and fasten the second length of molding to the pediments and fluted decorations in the same way.

5. Placing the head assembly.
◆ Butt the head unit against the top of the two pilasters, overlapping them by equal amounts at each end *(left)*.
◆ Fasten the assembly to the plywood base and the top of the pilasters with glue and four nails driven through each of the pediments and fluted decorations.

6. Adding the crown backer.
◆ On a table saw, rip a 1-by-3 $1\frac{3}{4}$ inches wide, then cut it $5\frac{1}{2}$ inches shorter than the length of the top base piece to make the crown backer (in this case, $50\frac{3}{4}$ inches long).
◆ With the backer flat on the plywood, butt it up against the head assembly, centering it along the width of the top base piece, and fasten it to the plywood base with glue and nails driven every foot *(right)*.

7. Adding the top cap.
◆ Cut a 1-by-4 to the length of the top base piece plus $\frac{3}{4}$ inch to allow for the picture-frame molding along the outer edges *(page 89, Step 5)*.
◆ Set the cap on edge against the top base piece.
◆ Drill pilot holes for countersunk $1\frac{1}{2}$-inch No. 8 galvanized wood screws 4 inches from each end and every foot in between.
◆ Fasten the cap to the edge of the base piece with glue and screws *(left)*.

8. Fastening the crown molding.
◆ With a manual or power miter saw *(page 21)*, bevel both ends of a length of $2\frac{1}{2}$-inch crown molding so the long edge is $\frac{3}{4}$ inch shorter than the top base piece ($53\frac{1}{2}$ inches in this case).
◆ Set the molding in place on the crown backer, butting it up against the head assembly and the top cap and centered on the unit end-to-end. Fasten the molding to the backer and top cap with glue and $1\frac{1}{2}$-inch galvanized finishing nails *(right, top)*.
◆ Fashion a return piece by making a compound cut at the end of another short length of molding and cutting it off square so its long edge is $2\frac{3}{4}$ inches long.
◆ Slip the mitered return against the end of the crown molding *(right, bottom)* and glue it to the top cap and plywood base; fasten it to the adjoining molding with glue and two 1-inch nails.
◆ Cut and fasten a return at the opposite end in the same way.

BUILDING THE ROOF UNIT

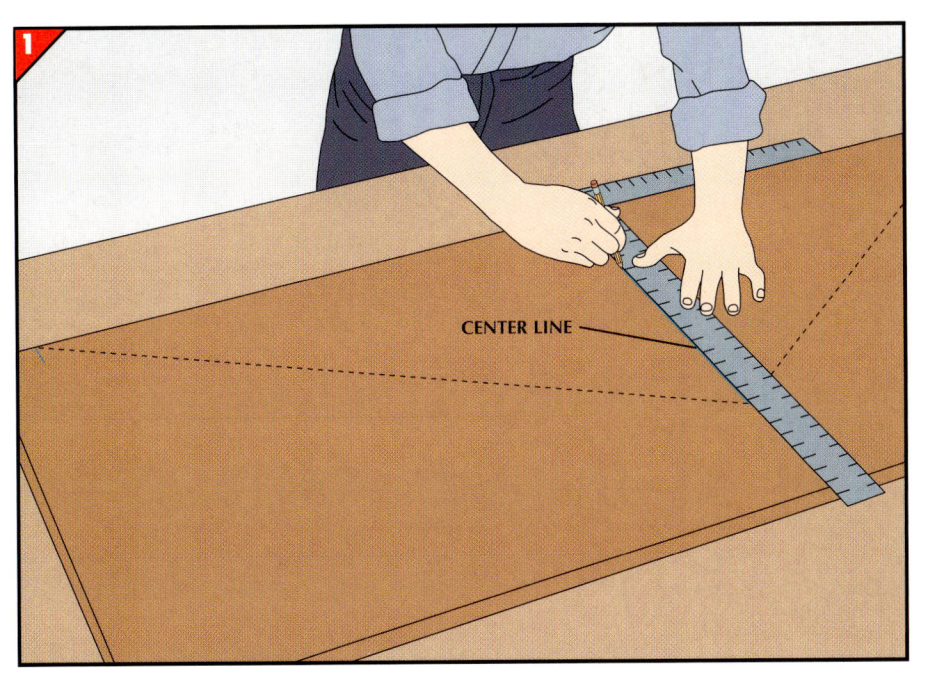

1. Cutting the triangular base.
◆ On the straight edge of a piece of ¾-inch exterior-grade plywood, mark the width of the base unit—here, 56¼ inches.
◆ Find the center of this measurement and make a mark there.
◆ Using a carpenter's square, measure up from the center point to a distance one-quarter to one-third the width of the base unit and draw a line *(left)*.
◆ Join the marks to form a triangle *(dashed lines)*.
◆ Cut out the triangle with a circular saw, using an edge guide for accuracy.

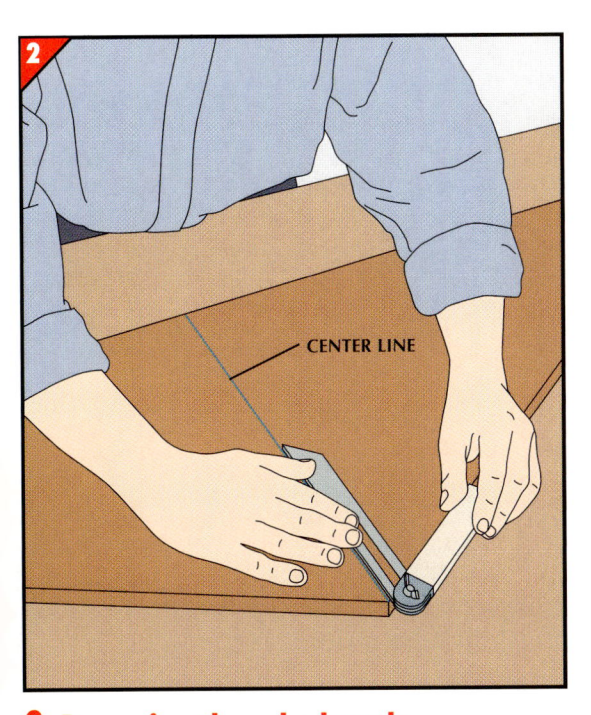

2. Preparing the rake boards.
◆ Set an adjustable T-bevel to the angle between one of the sloping sides of the triangle and the center line *(above)*.
◆ Adjust the blade of a manual or power miter saw to the angle and bevel one end of a 1-by-4 to make a rake board.
◆ Cut the other end of the board square so it is about 1½ inches longer than the sloping edge of the triangular base.
◆ Cut a second rake board for the opposite side of the roof in the same way.

3. Attaching the rake boards.
◆ Set one rake board against the sloping edge of the triangular base, aligning the angled end with the peak of the triangle. Drill countersunk pilot holes for 2-inch No. 8 galvanized wood screws 2 inches from each end of the sloping edge and every foot in between. Fasten the rake board to the base with glue and screws.
◆ Apply glue to the angled ends of both rake boards and install the second rake board in the same way as the first.
◆ Drive two 2-inch galvanized finishing nails through each side of the top joint, staggering the holes so the nails won't strike each other *(above)*.

4. Cutting the crown molding.

♦ With the miter saw adjusted to the angle measured in Step 2, make a compound cut *(page 21)* on one end of two lengths of crown molding.
♦ Mark the other end of the molding so the long edge of each piece matches the sloping sides of the base.
♦ Cut a straight 2-by-4 about 8 inches long and fasten it along one edge of an 8-inch-square piece of $\frac{3}{4}$-inch plywood. Clamp this jig to the base of the saw so the 2-by-4 faces the saw blade and is at right angles to the fence.
♦ Adjust a T-bevel to the angle at the bottom corner of the triangular base and adjust the saw blade to this angle relative to the jig's 2-by-4.
♦ Holding the crown molding securely in position against the jig, make the cut *(above). (In the illustration, the blade guard is removed for clarity)*. Cut the second piece in the same way.

5. Attaching the pieces.

♦ Fit one length of crown molding into position inside the rake boards. Fasten it to the triangular base and the rake board with glue and $1\frac{1}{2}$-inch galvanized finishing nails spaced every 8 to 10 inches *(above)*.
♦ Apply glue to the mating ends of the molding pieces and fit the second piece into position, fastening it with glue and nails.
♦ If necessary, use medium-grade sandpaper to sand the exposed ends of the crown molding flush with the bottom of the triangular base.
♦ Sink all the nails with a nail set and fill the holes with spackling compound.

A TRADITIONAL SUNBURST DECORATION

To give a completely different look to your entryway design, you can substitute a sunburst for the peaked roof. Available in polyurethane to withstand the elements, the unit is simply glued in place with a silicone-base exterior adhesive and is painted along with the rest of the entryway. If you plan to include this element in your design, omit the fluted vertical pieces that are part of the head assembly.

INSTALLING THE ENTRYWAY

1. Building out the jambs.
◆ For a brick-sided house, measure the distance from the front edge of the jamb to the front face of the exterior wall.
◆ Rip 1-inch lumber to this width on a table saw and cut the pieces to length to fit flush with the door jambs and header.
◆ Fasten the side pieces to the top piece with glue and countersunk 2-inch No. 8 galvanized wood screws, forming a jamb extension.
◆ Tilt the extension into place *(right)* and fasten it to the jambs with glue and galvanized finishing nails long enough to penetrate the jambs by $1\frac{1}{2}$ inches.

For a house with siding, prepare the wall as described in the box on page 88.

2. Setting the surround in place.
◆ Remove the brace that you installed temporarily *(page 81, Step 1)* to hold the unit square.
◆ Lift the door surround into position and have a helper steady it.
◆ Move the unit back and forth sideways until it is centered around the door opening, leaving a $\frac{3}{8}$-inch reveal on both sides of the jamb extension to allow for the edge molding.
◆ Raising the unit with a pry bar, slip shims under one side base piece *(left)*, then the other, until the top of the unit parallels the top of the jamb extension with a $\frac{3}{8}$-inch reveal.

DEALING WITH SIDING

If your house is covered with wood or vinyl siding instead of brick, you will need to cut away some of the siding to accommodate a door surround. First, outline the unit on the exterior wall, then cut around the outline with a circular saw set to the thickness of the siding and pry away the siding. Cut four lengths of J-channel—special flashing that keeps water from running behind the siding—to line the perimeter of the opening, apply exterior caulk to the outside of the long legs of the channels, and slip the pieces into place between the siding and the building paper. Fasten the surround to wall studs behind the sheathing and run a bead of caulk between the unit and the J-channel.

CAUTION: *Before cutting into wood siding, check for lead paint with a home test kit. If it is present, inquire at your local health department on safe practices for removing siding covered with lead paint.*

J-CHANNEL

3. Attaching the surround.

◆ Drill shallow counterbored holes for 2-inch masonry screws every 2 feet or so into the plywood base on each side piece of the unit *(left)*, positioning each hole so it aligns with the center of a brick.
◆ On the top piece of the unit, drill a hole at each top corner and two holes along the bottom edge, aligning the holes with brick centers.
◆ Fit the drill with a masonry bit and make a pilot hole into the brick through each hole.
◆ Drive a masonry screw into each hole.

4. Adding the roof.

◆ Squeeze a bead of exterior caulk along the back top edge of the top cap.
◆ Set the roof unit in place, centering it over the top cap *(left)*.
◆ Fasten the roof unit to the brick with four masonry screws—one on each side of the peak and one at each of the lower corners—in the same way you fastened the door surround *(Step 3)*.
◆ Cover all the screwheads with spackling compound.

5. Finishing the edges.

◆ Attach $\frac{3}{8}$-inch picture-frame trim to the inside edges of the plywood base as for a fireplace surround *(page 41)*.
◆ For the outside, cut two lengths of the trim long enough to cover the exposed outer edges of the plywood base.
◆ Fasten the trim to each outside edge of the plywood with glue and $1\frac{1}{4}$-inch galvanized finishing nails, spaced 2 inches from each end and every 8 to 10 inches in between *(right)*.
◆ Remove the shims under the unit.
◆ Run a bead of caulk all the way around the unit where it meets the brickwork and along the base of the unit, but leave a small gap on each side to allow water to escape.
◆ Sink all the nails with a nail set and cover the heads with spackling compound.

4 Crafting a Staircase

A well-built, sturdy staircase not only provides a passageway between floors, it also can add beauty and charm to a house. A simple straight-run design can replace an ailing main staircase or transform a set of utility stairs into the centerpiece of a basement renovation. Inclusion of a landing and a 90-degree turn will allow you to fit the structure in almost any space.

Planning and Design   92
Calculating Dimensions

Constructing a Solid Landing   98
Building the Structure

Fitting the Stringers   101
Marking the Notches
Hanging the Stringers

Cutting and Setting Newel Posts   108
Notching the Bottom Member
Fitting the Landing Newel
Attaching the Top Post

Adding Skirts, Risers, and Treads   114
Making the Wall Skirt
Attaching the Outer Sections
Installing Treads and Risers

Putting in Handrails and Balusters   120
Installing the Railing
Fitting the Balusters
Adding a Floating Handhold

Attaching a riser →

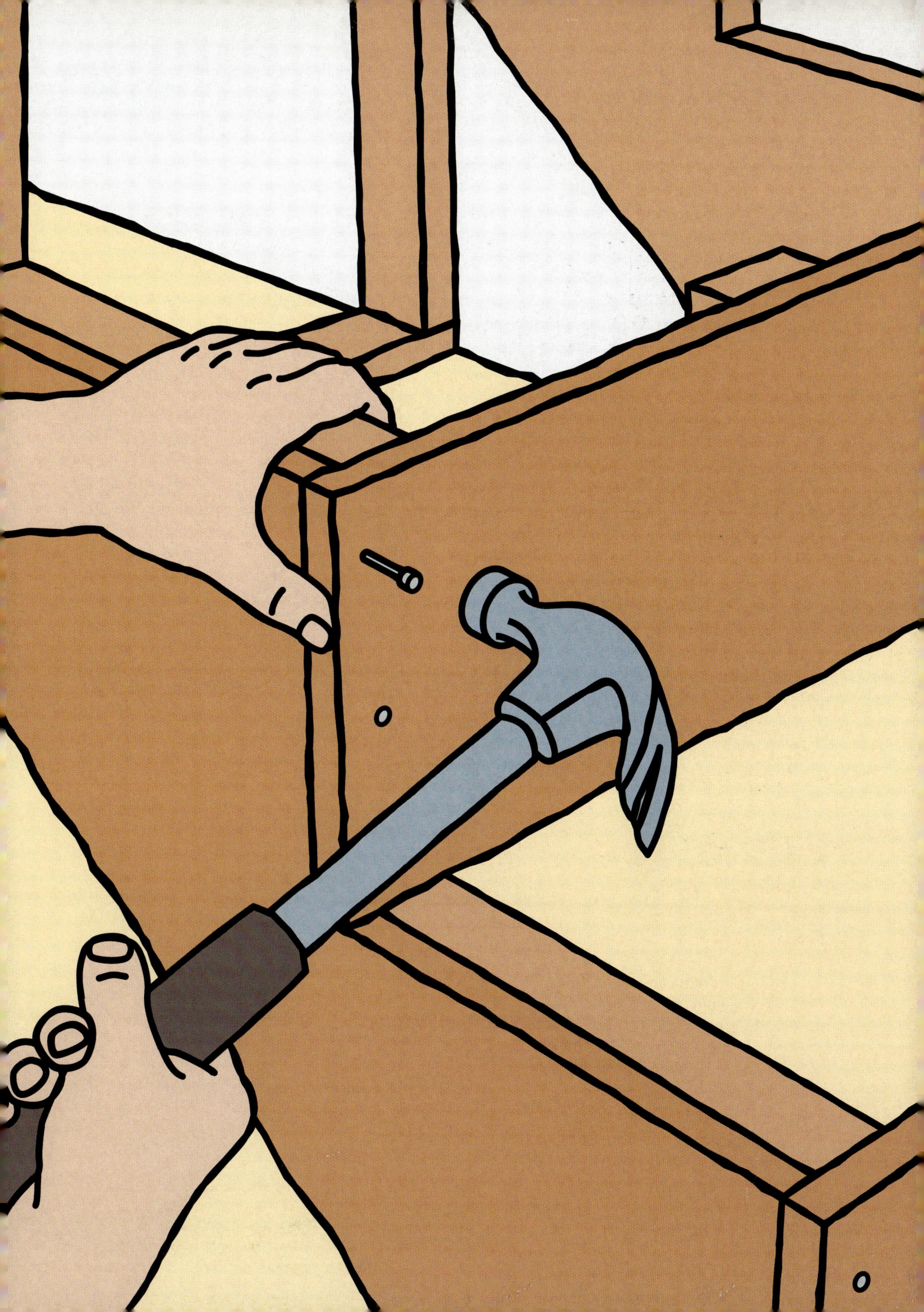

Planning and Design

Simple or complex, a staircase requires careful planning to fit the available space and to provide a safe, convenient passageway between floors. The structure on these pages has an uncomplicated design, but is embellished with an elegant railing and an attractive hardwood finish. A set of utility stairs can be built on the same principles without the ornamental touches.

Planning the Job: For the stairs to be safe and accessible, you will need to follow the building codes that govern many aspects of the construction, such as the angle of the stringers and the height of the handrails *(page 94)*. The most complex part of the process is figuring the relative dimensions of the various components *(pages 95-96)*.

If you add a set of stairs where there was none, you may need to frame a new opening in the floor above the stairs *(page 97)*.

Adding a Landing: To fit a flight of stairs into a limited space, you can incorporate a landing. Depending on the floor plan of the house, you can choose among several configurations *(right)*. In the structure shown on the following pages, the landing falls in the middle of the flight, but you can locate it closer to the top or bottom if desired *(page 96)*, or omit it entirely.

Three designs.
Stairs built with a landing can make a 90-degree turn in either direction, creating a simple L shape *(right, top)*. Attaching an upper flight to opposite sides of a landing creates a T-shaped staircase, providing access to separate areas of the upper floor *(right, center)*. Enlarging the landing and attaching two flights to the same side allows a 180-degree turn where the stairs double back on themselves *(right, bottom)*.

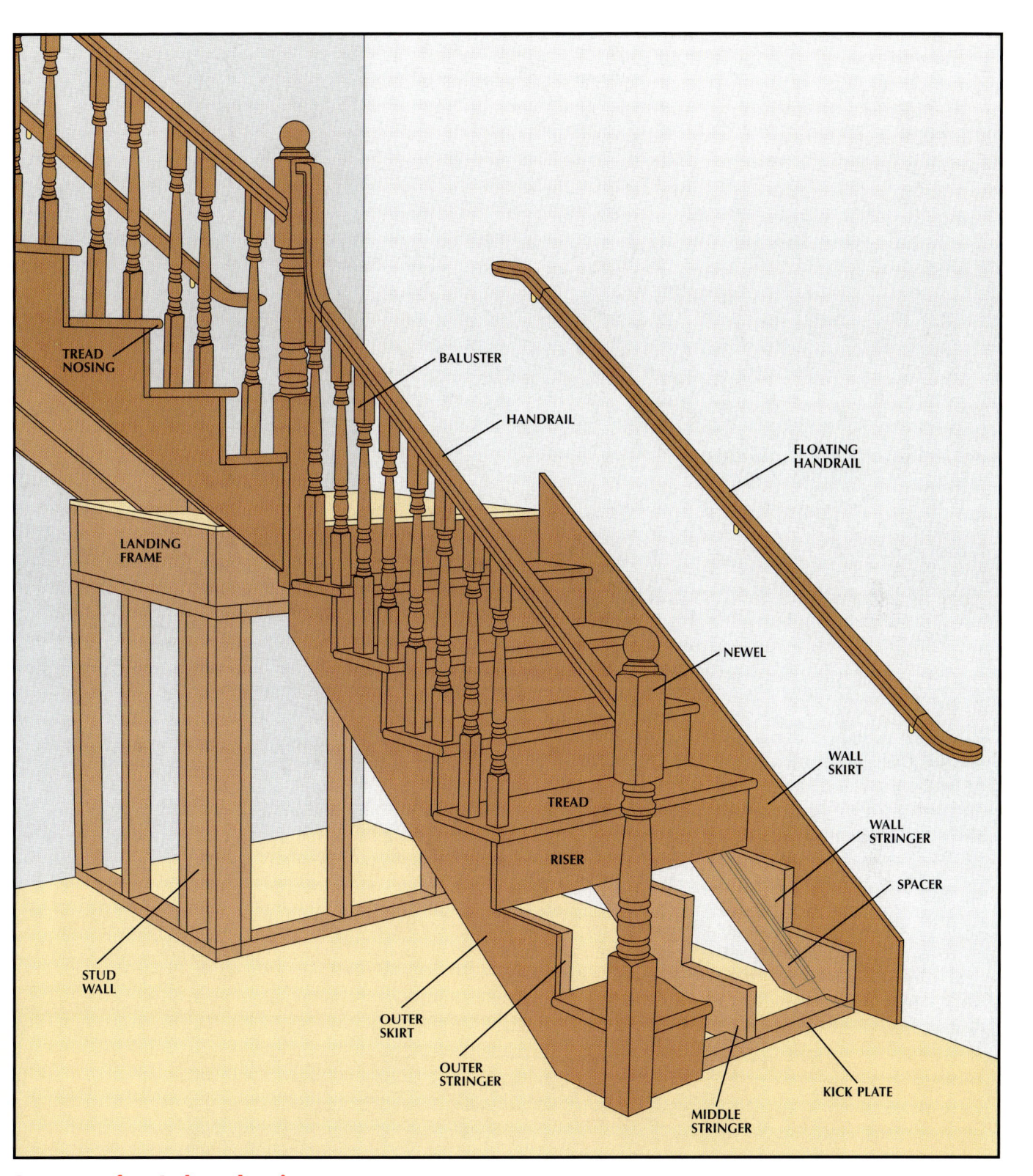

Anatomy of an L-shaped staircase.

This staircase makes a 90-degree turn at a landing halfway up. The landing is a simple frame attached to the corner walls and supported by two outer stud walls. Three notched stringers form the basis of the steps, and hardwood treads and risers are screwed to them. The stringers of the lower flight are fastened to a hangerboard secured to the landing *(page 104)* and are braced at the bottom by a kick plate; the upper flight is also stabilized by a kick plate on the landing, and is secured to a hangerboard on the framing of the upper floor. Hardwood skirts protect the wall from scuffing and hide the rough lumber of the outer stringers. Newel posts support a handrail and balusters and, for added safety, a floating handrail runs along the wall.

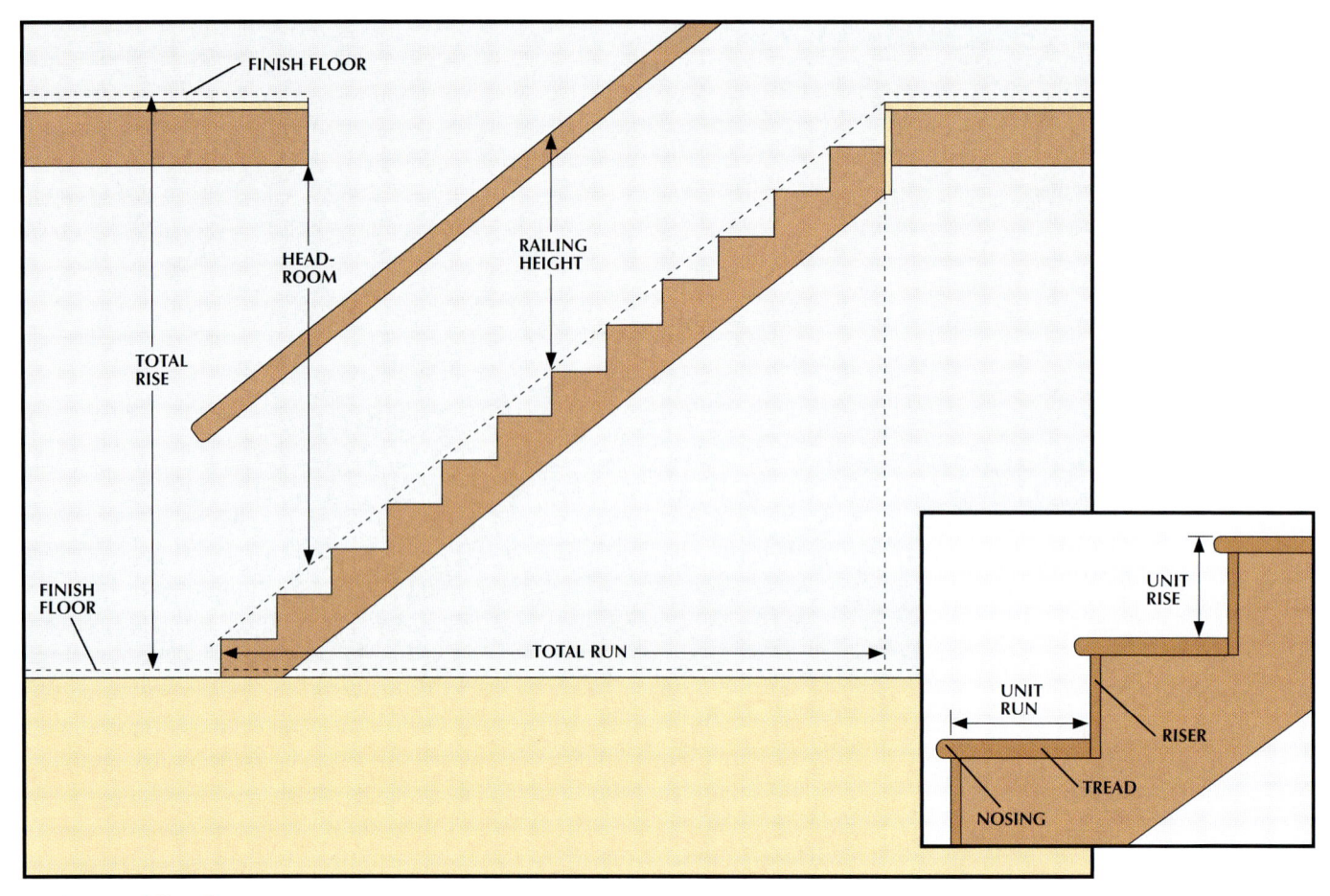

Stair specifications.

In planning a staircase, you need to consider the relationships between all of its parts. The distance the stairs cover vertically—the total rise—is dictated by the space between floors; the horizontal distance—the total run—may be limited by existing space; but building codes, whether national or local, govern the specifications of most of the other elements. According to most codes, a staircase's minimum width, from the finish wall to the center of the railing (or to an opposite finish wall), must measure at least 36 inches—but 40 inches is a more comfortable width if you have the space. Headroom—the vertical distance between the ceiling and an imaginary diagonal line connecting the noses of all the treads—is usually 80 to 84 inches minimum. A formula (below) governs the relationship between the vertical distance between treads—the unit rise—and the horizontal distance between risers—the unit run (inset).

The part of the tread that overhangs the riser—called the nosing—is generally $\frac{3}{4}$ to $1\frac{1}{4}$ inches deep, but the actual overhang depends partially on the width of the tread stock. If you use $11\frac{1}{4}$-inch stock for 10-inch treads, the nosings will be $1\frac{1}{4}$ inches deep. In some localities, staircases with a unit run of more than 11 inches do not require a nosing. Handrails are typically 30 to 34 inches above the stairs, as measured from the nosings to the top of the rail.

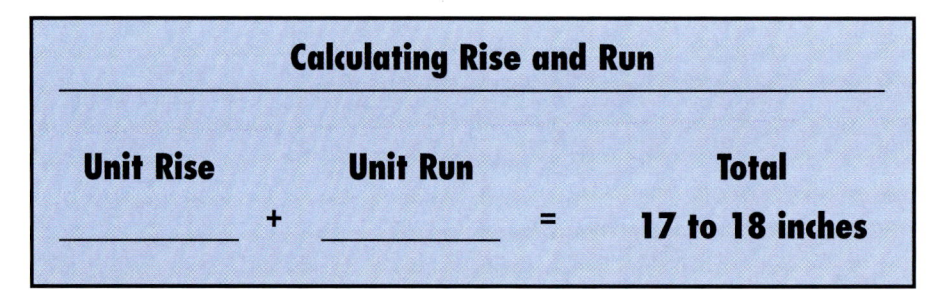

Calculating Rise and Run

Unit Rise	Unit Run	Total
_____ +	_____ =	17 to 18 inches

Restrictions on unit rise and run.

To ensure the safety and convenience of a staircase, codes generally allow a maximum unit rise of about $7\frac{3}{4}$ inches and a minimum run of 9 to 10 inches. The relationship between the unit rise and the unit run is governed by the formula shown above: The combined rise and run must fall between 17 and 18 inches. A rise of $7\frac{1}{2}$ inches and a run of 10 inches would be an ideal combination. All steps on every flight of the staircase must have the same unit rise and run.

CALCULATING DIMENSIONS

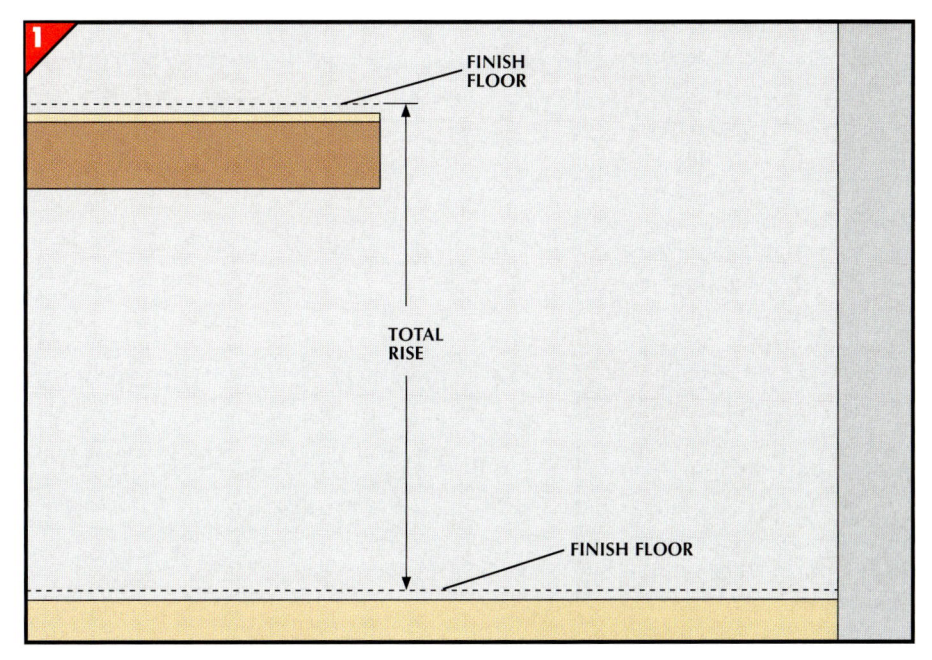

1. Determining unit rise and run.

◆ Measure the total rise from the lower finish floor to the upper one. If the finish floors are not in place, measure from subfloor to subfloor, subtract the intended thickness of the lower floor material, and add the thickness of the top floor material *(above)*.
◆ Divide this figure (105 inches, for example) by the maximum allowable rise ($7\frac{3}{4}$ inches) to obtain a tentative number of risers ($13\frac{1}{2}$). Round this result up to the nearest whole number (14).
◆ Divide the total rise (105 inches) by the number of risers (14) to get the actual unit rise ($7\frac{1}{2}$ inches).
◆ To find the ideal run, use the formula opposite. From the result ($9\frac{1}{2}$ to $10\frac{1}{2}$ inches), select a whole number within the range (10 inches).

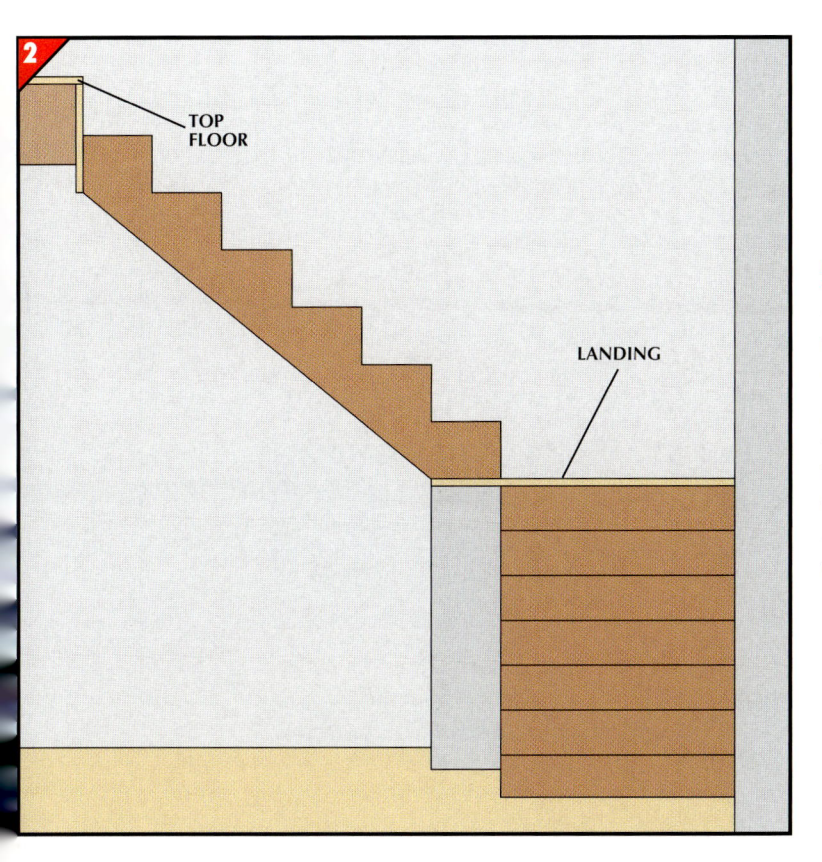

2. Totaling the risers and treads.

◆ For a landing halfway between flights, divide the total number of risers (14) by two to get the number of risers per flight (seven).
◆ Because the landing and top floor form the top tread of each flight, subtract one from the number of risers to determine the number of treads per flight (six).

3. Checking the fit of the stairs.

To determine if the proposed stairway will fit in the existing space, make some simple calculations.

◆ First, find the horizontal space required for each flight. Calculate the total run of each flight by multiplying the number of treads (six) by the ideal unit run (10 inches). Add to the resulting figure (60 inches) the width of the stair structure *(page 98)*—(in this staircase, $40\frac{3}{4}$ inches); then add $\frac{3}{4}$ inch for the thickness of the hangerboard at the top and $\frac{3}{4}$ inch for the thickness of the riser at the bottom ($102\frac{1}{4}$ inches).

◆ Measure the length of the opening in the upper floor, then the available space on the lower floor to see if each flight will fit. If either measurement is less than the horizontal space needed ($102\frac{1}{4}$ inches), reduce the unit run for both flights slightly or adjust the height of the landing *(box, below)*. If the upper-floor opening is too large, you can increase the unit run for both flights slightly or add a riser, provided you respect the formula on page 94 and recalculate the stair dimensions.

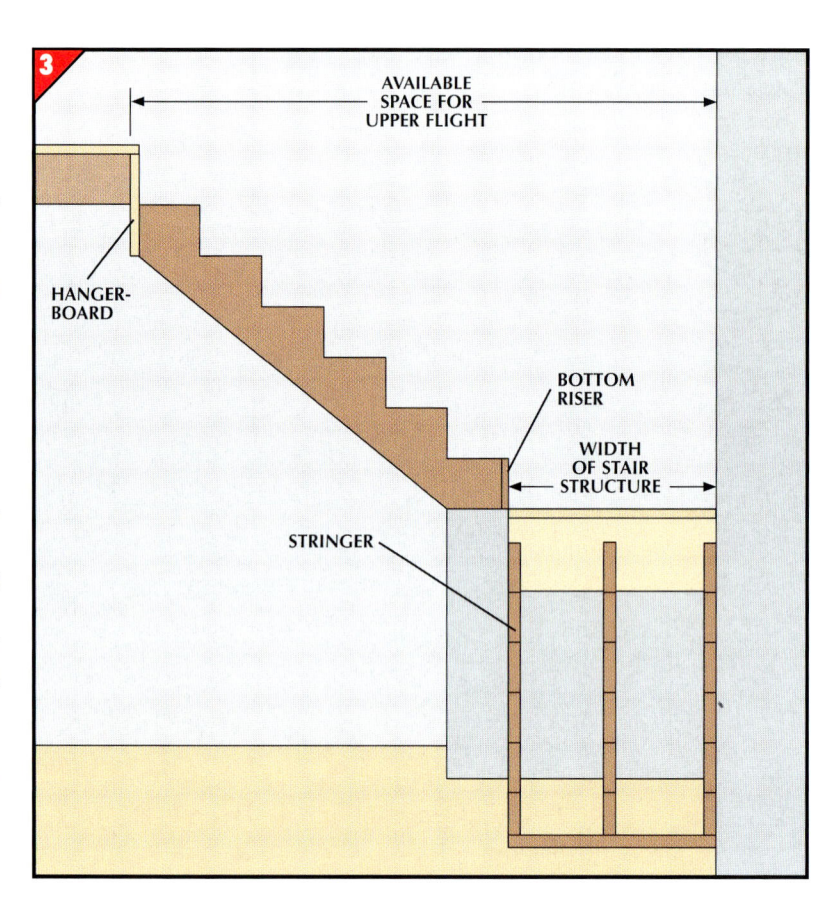

ADJUSTING FOR LIMITED SPACE

Placing a landing halfway between flights simplifies the planning and building process, but since the two flights will occupy the same amount of room, this design may not be suited to every situation. By lowering or raising the position of the landing *(below)*, you can adapt the configuration of the stairs to fit the available space on the lower floor.

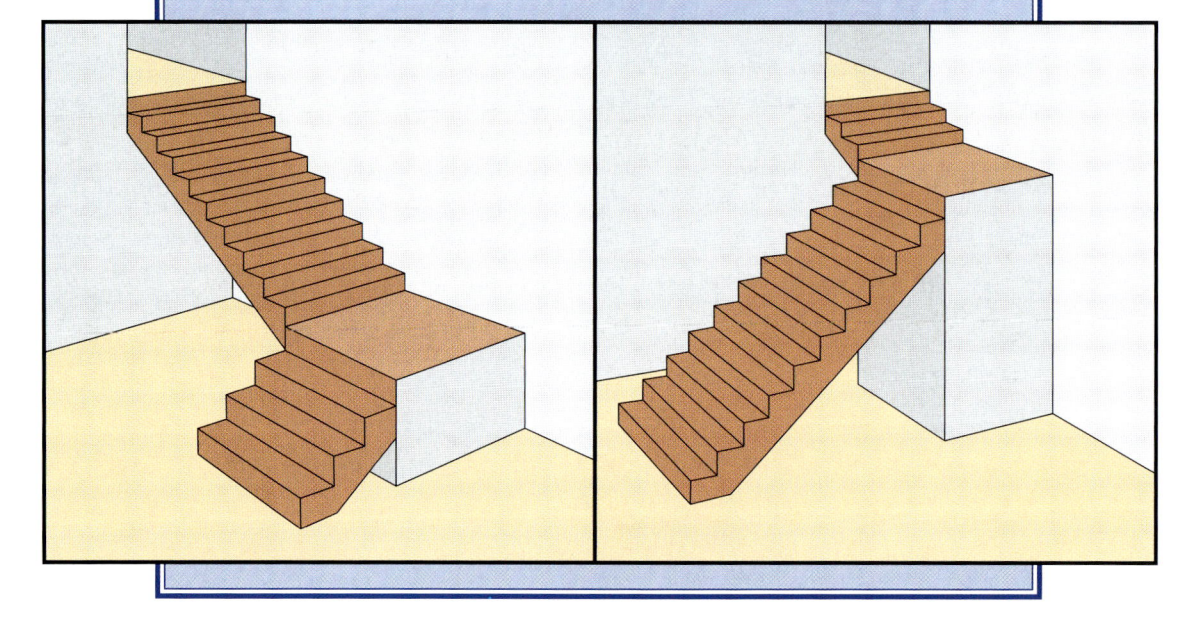

REINFORCEMENT FOR A STAIR OPENING

When a staircase is built, an opening must often be cut and framed in the floor above it. The size of the opening depends on the stairs' shape and on the location of the landing, and is designed to afford at least the minimum required headroom at every point along the stairs *(page 94)*. A straight flight with no landing needs a long, narrow opening, while a stairway with a landing halfway up needs a shorter almost square opening.

Before the opening is made, temporary walls are constructed, near the point where the joists will be cut, to hold up the floor until the framing is finished. (If one side of the opening is at a bearing wall, as shown below, it does not need a temporary wall there.) Next, "trimmer" joists are are added alongside the planned opening for extra strength, and are stabilized with blocking. Then, the floor joists are cut and headers of doubled joist lumber nailed to the ends of the cut joists—called tail joists—and to the trimmers; when the opening extends to a bearing wall, a single header is nailed to the ends of the joists sitting on the wall.

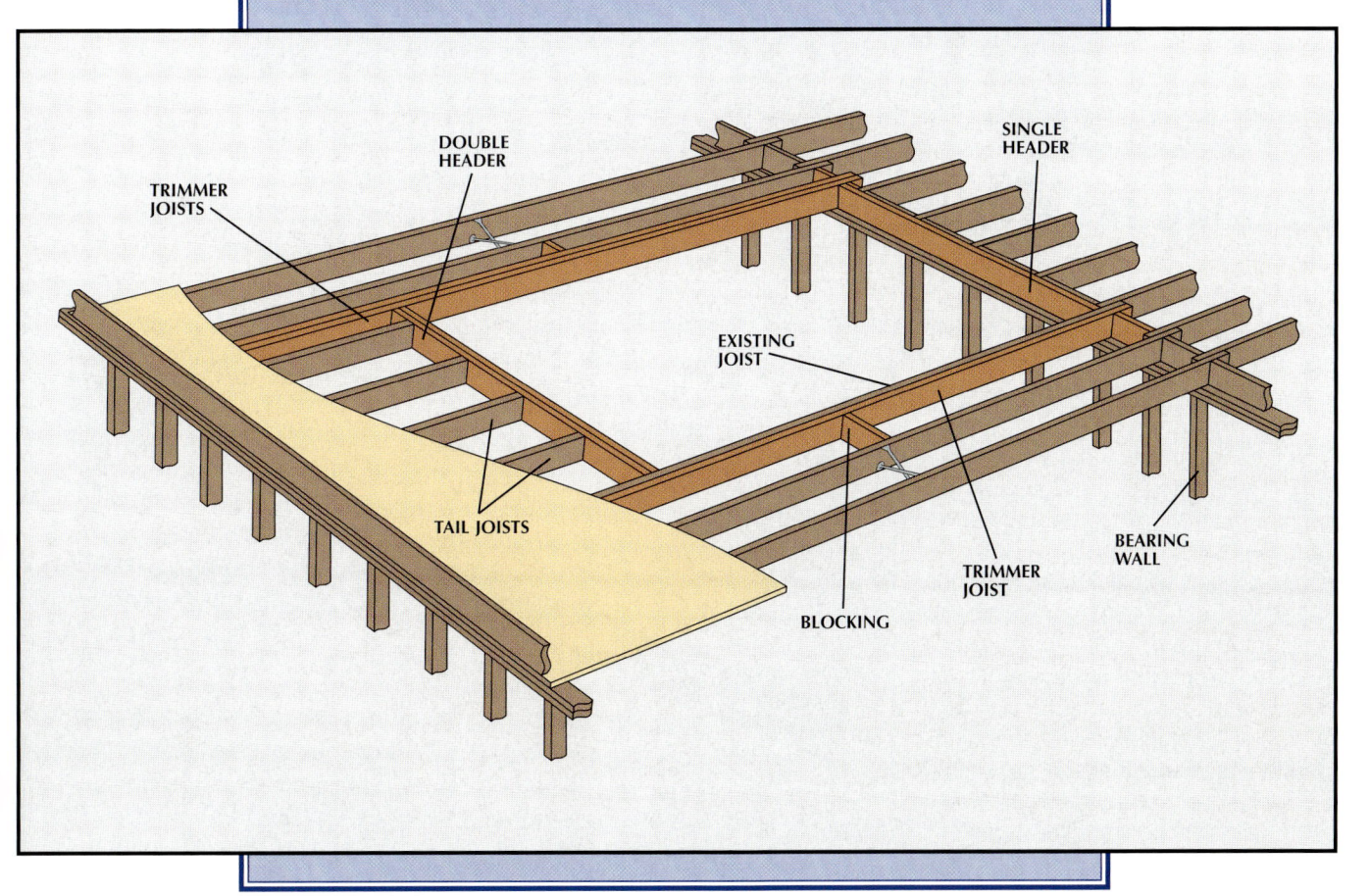

Constructing a Solid Landing

The landing of an L-shaped staircase can be located at any height, depending on the available space *(page 95)*. In the simplest design, it is halfway between the upper and lower flights.

Figuring the Dimensions: Establish the landing's height *(below)*, then calculate its width and depth as determined by the width of the stair structure. To simplify the building process, use whole lengths of 42-inch tread stock and make the width of the stair structure—as measured from the wall to the outside edge of the outer stringer—$40\frac{3}{4}$ inches (42 inches minus $1\frac{1}{4}$ inches for the treads' overhang at the balusters). Use this figure for the landing's depth—measured from the top of the lower flight of stairs to the back wall. To figure the landing's width—starting at the side wall and extending beneath the stringers of the upper flight—add 10 inches to accommodate the feet of the upper stringers and $\frac{3}{4}$ inch for the thickness of the bottom riser of the upper flight. For the stairs on these pages, the landing would measure $40\frac{3}{4}$ by $51\frac{1}{2}$ inches, but you can adapt the dimensions to suit your situation.

TOOLS
Tape measure
Carpenter's level
Electronic stud finder
Circular saw
Hammer
Electric drill
Screwdriver bit
Utility knife

MATERIALS
2 x 4s, 2 x 10s
Plywood ($\frac{3}{4}$")
Shims
Common nails ($3\frac{1}{2}$")
Wood screws ($3\frac{1}{2}$" No. 12)
Flooring screws ($1\frac{3}{4}$")
Construction adhesive

SAFETY TIPS
When driving nails or using a power tool, protect your eyes with goggles.

BUILDING THE STRUCTURE

1. Marking the landing location.
♦ Measure from the floor to the desired height of the landing, subtracting the thickness of the landing's planned finish floor and adding the thickness of the lower finish floor if it is not yet in place. Make a mark on one wall at this height to indicate the top of the landing subfloor.
♦ Measure down $\frac{3}{4}$ inch from this spot and make another mark to indicate the top of the framing.
♦ At the lower mark, use a carpenter's level to draw a line along the wall for the top of the framing *(right)*, then continue the line along the adjacent wall.
♦ With an electronic stud finder, locate and mark the stud positions on both walls.

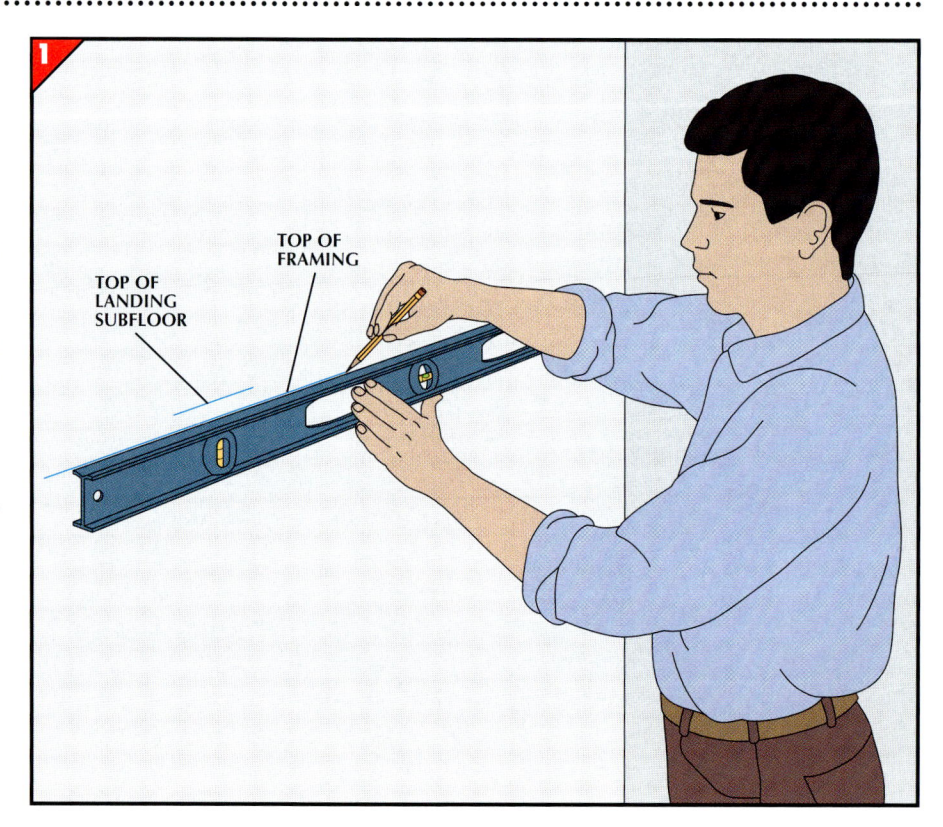

2. Erecting the first wall.
◆ Using the planned dimensions of the landing (opposite), build one 2-by-4 stud wall to reach from the floor to a point $9\frac{1}{2}$ inches below the framing line on the wall, fastening the studs to the top plate and soleplates on 16-inch centers with $3\frac{1}{2}$-inch common nails.
◆ With the level as a guide, mark a plumb line on the back wall to indicate the width of the landing; if the line does not run along a stud, cut into the back wall between the studs where the landing wall will be situated and add 2-by-4 blocking between the studs about every foot (inset).
◆ Driving the nails partway, tack the stud wall to the back wall so the outer edge of the end stud is aligned with the plumb line.
◆ Drive shims under the soleplate as necessary to level the wall (left).

3. Adding the second wall.
◆ Build a second stud wall the same height as the first to the planned width of the landing, less $3\frac{1}{2}$ inches to allow for the thickness of the studs in the first wall.
◆ Measure along the side wall a distance equal to the dimensions of the landing and mark a plumb line.
◆ Add blocking behind the wall if necessary then, with the outside of the second stud wall lined up with the plumb line and the end of the first stud wall, tack it to the side wall.
◆ Level the second stud wall with shims.
◆ Tack the second stud wall to the first with four nails (right).

4. Checking for square.
Measure the structure diagonally from corner to corner (left), then do the same across the other two corners. The stud walls are square if the measurements are identical. If they are not, try driving them into square with a hammer. If this does not work, remove the nails and reposition the walls.

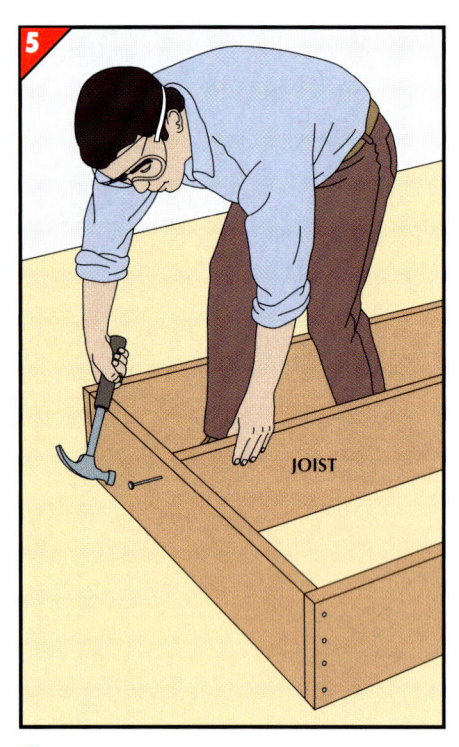

5. Assembling the frame.

◆ Cut two 2-by-10s to the width of the landing and two more to its depth less 3 inches.
◆ Nail them together in a rectangle, sandwiching the shorter boards between the longer ones; drive four $3\frac{1}{2}$-inch nails at each joint.
◆ Cut two 2-by-10 joists to the width of the landing less 3 inches and nail them to the frame on 16-inch centers *(above)*.

6. Installing the frame.

◆ Working with a helper, set the frame onto the stud walls *(above)*.
◆ Check the frame for level in both directions and adjust the shims beneath the stud walls if necessary.
◆ Holding the top edge of the frame even with the location lines marked in Step 1, drill pilot holes for $3\frac{1}{2}$-inch No. 12 wood screws through the frame and into the wall studs. Drive in the screws.
◆ Toenail the frame to the top plates of the stud walls with $3\frac{1}{2}$-inch nails.
◆ Drive home the nails pinning the stud walls to each other and to the back and side walls.
◆ Fasten the soleplates of the stud walls to the subfloor.
◆ Score the protruding shims with a utility knife and break them off. Apply construction adhesive to the ends of the shims to keep them from working loose.

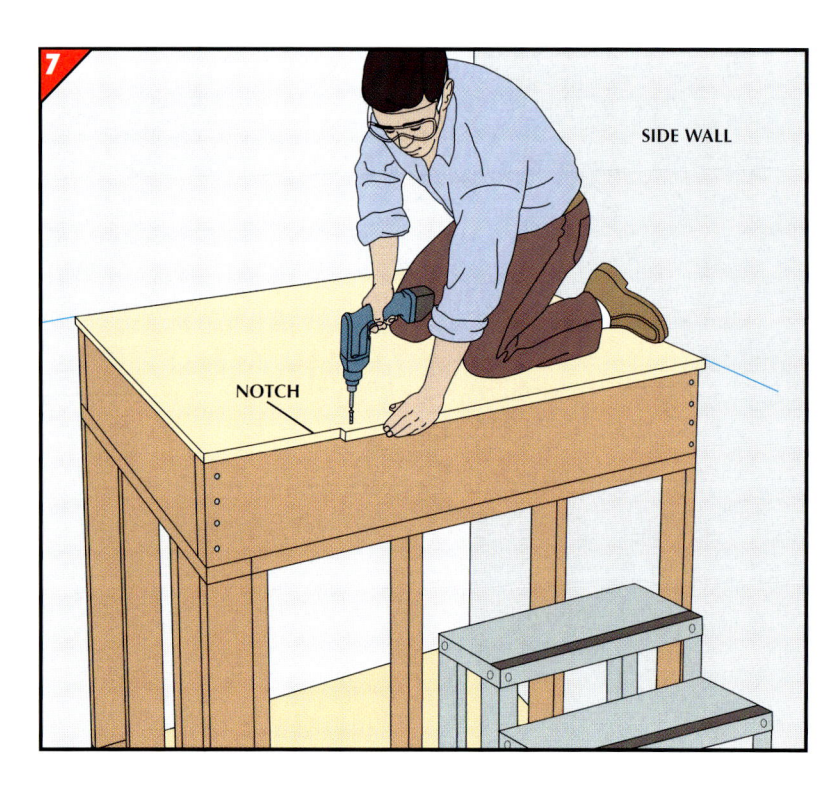

7. Adding the subfloor.

◆ Cut a piece of $\frac{3}{4}$-inch plywood as a subfloor to the width of the landing and $\frac{3}{4}$ inch more than its depth. (This $\frac{3}{4}$-inch overhang will cover the hangerboard for the lower flight of stairs.)
◆ Position the subfloor on the frame and measure from the side wall a distance equal to the width of the stair structure *(page 98)*, less $1\frac{3}{8}$ inches to account for the amount that the landing newel post will overhang the stringer *(page 112)*. Mark the subfloor at this point.
◆ Remove the plywood. With a circular saw, cut a $\frac{3}{4}$-inch strip from the front edge, stopping at the marked point, to create a notch that will accommodate the landing newel post.
◆ Apply construction adhesive to the landing joists and frame, then reposition the subfloor.
◆ Fasten the subfloor to the joists and frame with $1\frac{3}{4}$-inch flooring screws spaced every 6 inches around the perimeter and every foot in between *(left)*.

Fitting the Stringers

Supporting the treads and risers of a staircase is an understructure of notched stringers. Each flight of steps requires three of them. For stairs that feature a landing halfway between flights, the upper and lower stringers are identical.

Preparing the Pieces: Cut to length from 2-by-12s *(pages 102-104)*, stringers must be as straight as possible to support the treads and risers properly. The lumber can have small, tight knots, but avoid boards with large, loose knots. When laying out the notches in the stringers *(page 102)*, try to locate any knots in the wood in the pieces to be cut out.

Installing the Stringers: Hangerboards at the landing and the upper floor support the top end of the stringers. At the bottom end, kick plates fastened to the lower floor and the landing keep the stringers from pushing forward. Spacers hold the wall stringers away from the wall to accommodate a skirt board that will be installed later. Once the wall and outside stringers are in position, the middle stringers are centered between them.

TOOLS
Sawhorses
Carpenter's square
Square gauges
C-clamps
Circular saw
Handsaw
Wood chisel
Carpenter's level
Electronic stud finder
Hammer
Maul
Electric drill
Screwdriver bit
Pry bar

MATERIALS
2 x 4s
2 x 12s
Plywood ($\frac{3}{4}$")
Shims
Masonry nails ($2\frac{1}{2}$")
Common nails ($2\frac{1}{2}$", 3)
Flooring screws ($1\frac{1}{4}$")
Wood screws (3" No. 10, $3\frac{1}{2}$" No. 12)
Wood glue

SAFETY TIPS
Put on goggles when driving nails or using a power tool.

Stringer layout.
Although a flight of stairs has one more riser than it has treads, a stringer is laid out with an equal number of treads and risers. This is because stringers are installed one riser height below the upper floor or landing. The top of the stringer is squared off to fit against the upper floor or landing and the bottom is squared off to sit on the landing or lower floor.

Since there is no tread at the bottom of a flight of stairs, the bottom riser would be too high if it were the same height as the others. To calculate the height of the bottom riser—a process known as dropping the stringer—subtract the tread thickness ($1\frac{1}{8}$ inches) from the unit rise ($7\frac{1}{2}$ inches). If the finish flooring is not yet in place at the bottom of the flight, add its thickness ($\frac{3}{4}$ inch) to this figure, resulting in a bottom riser with a height of $7\frac{1}{8}$ inches in this example *(inset)*.

MARKING THE NOTCHES

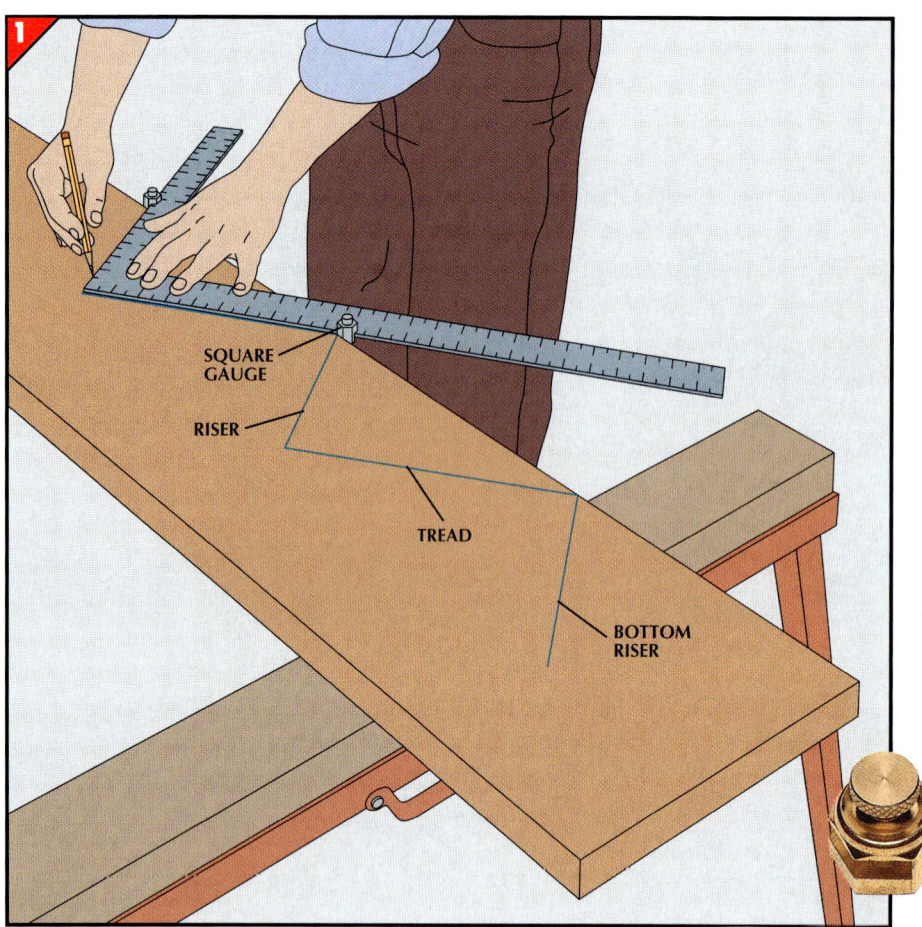

1. Laying out treads and risers.
◆ Lay a 2-by-12 of the appropriate length *(box, below)* across sawhorses.
◆ Attach two square gauges *(photograph)* to the outside edges of a carpenter's square, positioning one on the short arm at the inch mark for the unit rise and the other on the long arm at the inch mark for the unit run.
◆ Lay the square on the board so the gauges are in contact with the edge and mark the bottom riser.
◆ Slide the square along and mark a tread and a riser, starting the tread where the bottom riser ends.
◆ Continue marking the required number of treads and risers *(left)*; then mark the last tread.

TRICKS OF THE TRADE

Calculating Stringer Length

With a carpenter's square and a tape measure, you can quickly determine the length of 2-by-12s to buy for the stringers. In this technique, each inch mark on the tool represents 1 foot. Let the tape measure symbolize the stringer and place its end on the square's inch mark that corresponds to the total run measurement. Stretch the tape to the opposite arm of the square to the point that marks the total rise *(right)*; then, since the stringer is hung one riser below the landing, move the tape down one riser height below the total rise mark. Convert the resulting measurement on the tape from inches to feet to get the approximate stringer length, then add 2 feet for waste and buy lumber at least this length.

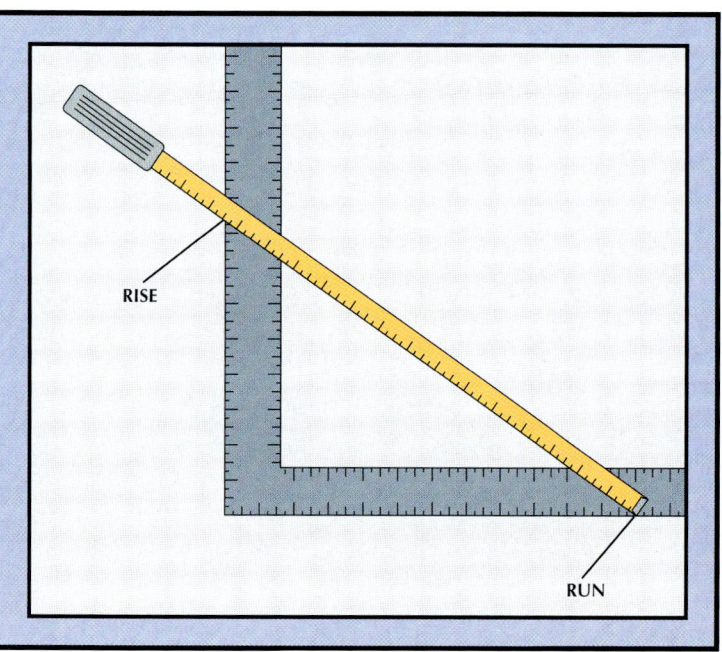

2. Completing the top.
◆ At the top of the layout, slide the square to the opposite side of the board without turning it around, making sure the gauges are in contact with the edge.
◆ Line up the short arm with the end of the top tread and mark a line at a right angle to it *(right)*.

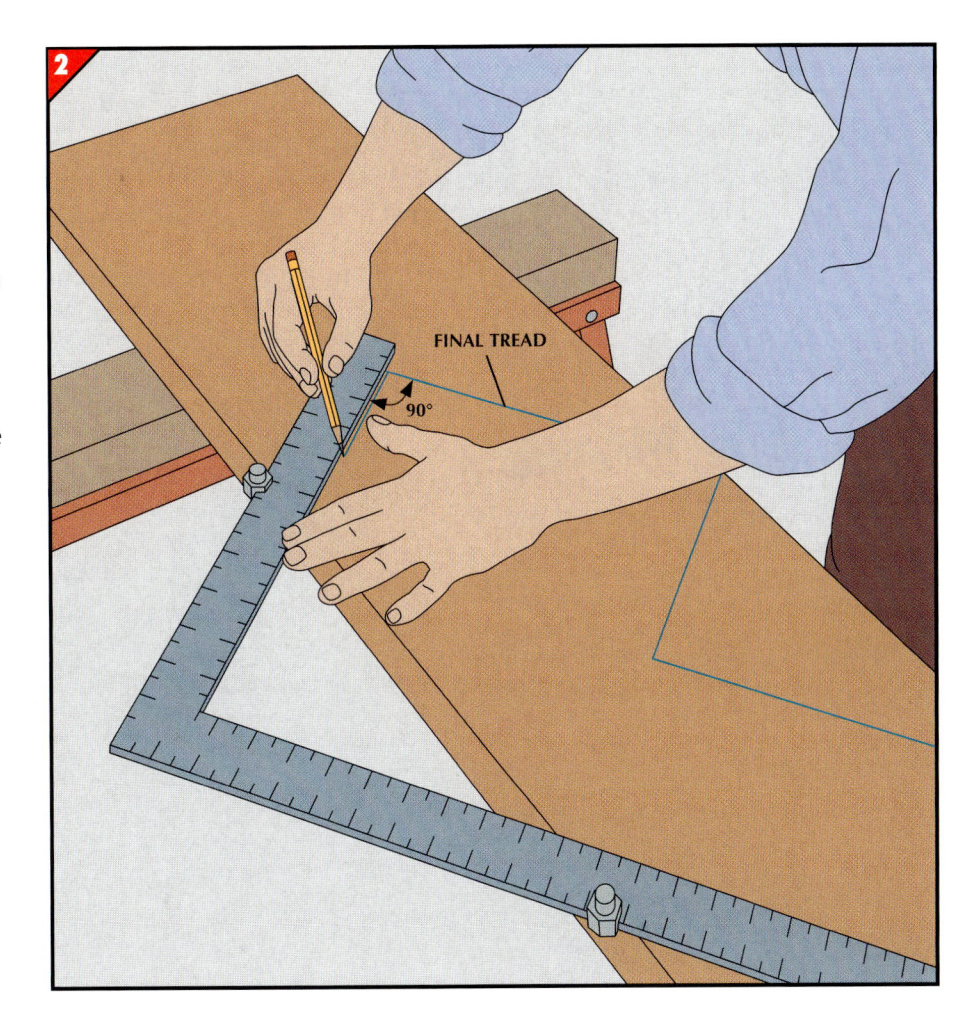

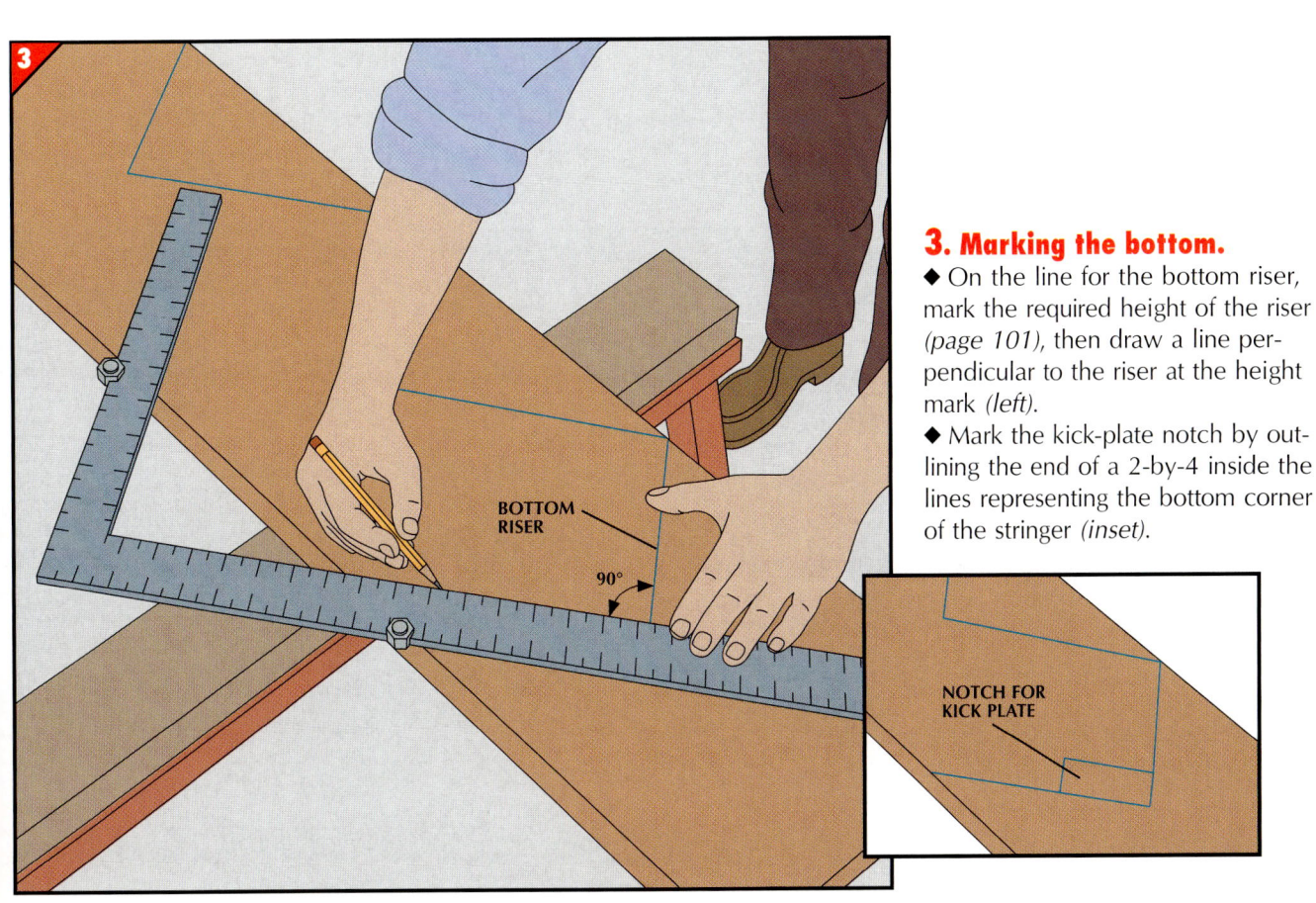

3. Marking the bottom.
◆ On the line for the bottom riser, mark the required height of the riser *(page 101)*, then draw a line perpendicular to the riser at the height mark *(left)*.
◆ Mark the kick-plate notch by outlining the end of a 2-by-4 inside the lines representing the bottom corner of the stringer *(inset)*.

4. Notching the stringers.

◆ Clamp the stringer board to the sawhorses.
◆ Cut along the waste side of the marked lines with a circular saw *(right)*.
◆ Complete the cuts with a handsaw. Smooth the cuts with a chisel, if necessary.
◆ Using the first stringer as a template, mark the remaining stringers for the upper and lower flights, then notch them.

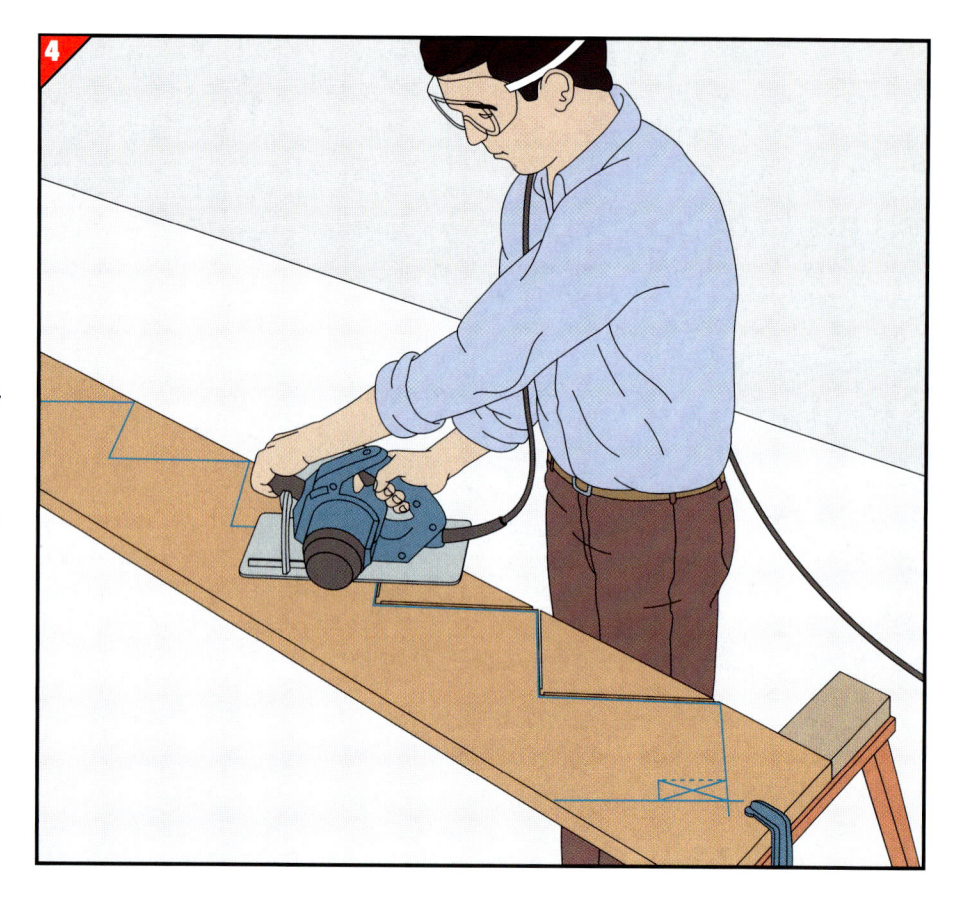

HANGING THE STRINGERS

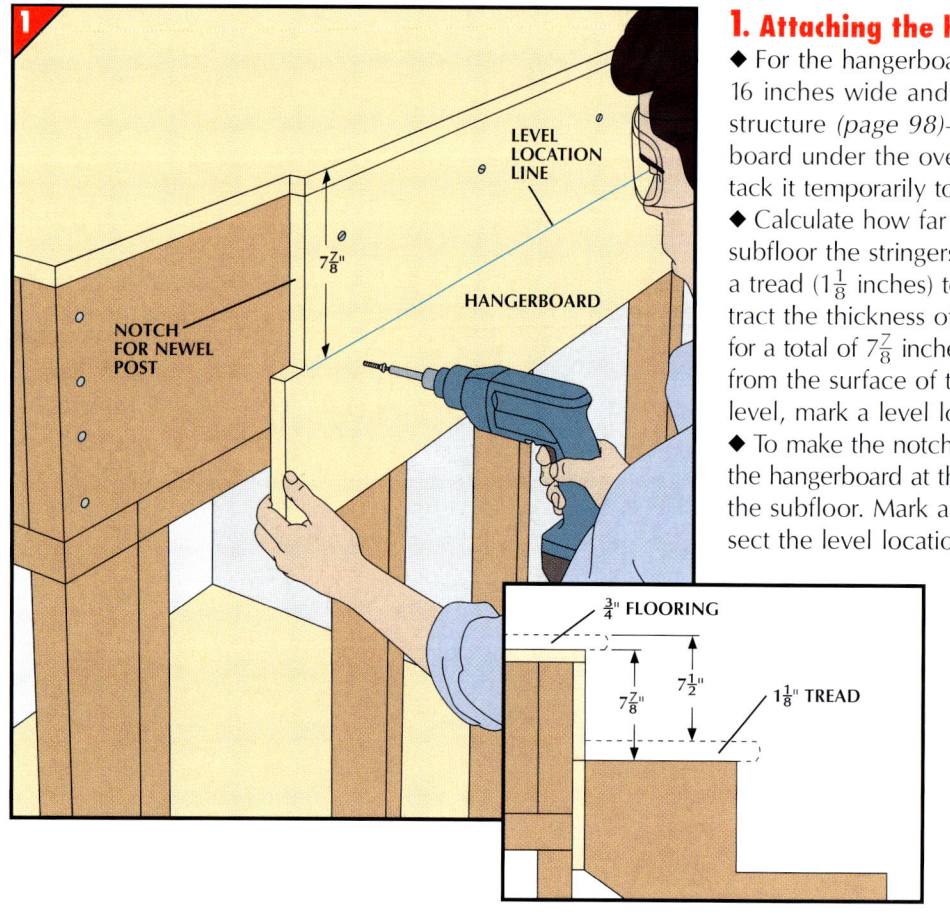

1. Attaching the hangerboard.

◆ For the hangerboard, cut a piece of $\frac{3}{4}$-inch plywood 16 inches wide and as long as the width of the stair structure *(page 98)*—$40\frac{3}{4}$ inches, in this case. Fit the board under the overhang of the landing subfloor and tack it temporarily to the landing frame.
◆ Calculate how far below the surface of the landing subfloor the stringers will hang: Add the thickness of a tread ($1\frac{1}{8}$ inches) to the unit rise ($7\frac{1}{2}$ inches) and subtract the thickness of the landing finish flooring ($\frac{3}{4}$ inch), for a total of $7\frac{7}{8}$ inches *(inset)*. Measure this distance down from the surface of the landing and, with a carpenter's level, mark a level location line on the hangerboard.
◆ To make the notch for the newel post, mark the top of the hangerboard at the point where it meets the notch in the subfloor. Mark a plumb line from this point to intersect the level location line. Remove the hangerboard and cut the notch for the newel post at the two marked lines.
◆ Apply wood glue to the face of the landing, then reposition the hangerboard. Drive in two rows of $1\frac{1}{4}$-inch flooring screws, spacing the screws every 6 inches *(left)*.
◆ Cut and install the hangerboard for the upper flight in the same way, fastening it to the framing of the upper floor.

2. Installing the wall spacer.
◆ Hold a stringer in place against the wall with its top level with the location line on the hangerboard and its bottom squarely on the floor. Trace a line on the wall along the lower edge of the stringer *(above, left)*.

◆ Remove the stringer and mark the center of each stud.
◆ Cut a 2-by-4 spacer to the length of the diagonal line.
◆ Holding the spacer with its lower edge along the line, nail it to the wall with two $3\frac{1}{2}$-inch nails driven into each stud *(above, right)*.

3. Tacking up the stringers.
◆ Measure along the hangerboard $1\frac{1}{2}$ inches from the spacer and make a mark, then with a carpenter's level, draw a plumb line at this spot.
◆ Set the wall stringer in place against the spacer, aligning it with the level location line and the plumb line. With two $2\frac{1}{2}$-inch nails, toenail the stringer to the hangerboard, driving in the nails only partway *(left)*.
◆ Starting at the wall, measure along the hangerboard the width of the stair structure and mark a plumb line.
◆ Hold the outer stringer in place, aligned with the level location line and the plumb line. Toenail the stringer to the hangerboard, driving in the nails partway.

4. Tacking down the kick plate.
◆ Cut a 2-by-4 kick plate to the width of the stair structure *(page 98)* less $1\frac{1}{2}$ inches to account for the wall spacer.
◆ Fit the kick plate into the notches at the base of the stringers and fasten it to the floor with four $2\frac{1}{2}$-inch common nails, driving them in partway *(left)*; for a concrete floor, drive in $2\frac{1}{2}$-inch masonry nails partway with a maul.

5. Leveling the stringers.
Set a level across the two stringers. If necessary, drive a pair of shims under the heel of the lower stringer to level them *(right)*. To fill a gap between the kick plate and the stringer, lift up the kick plate with a pry bar and drive shims under it.

6. Attaching the outer stringers.

◆ Nail the wall stringer to the spacer, driving a 3-inch nail every foot.
◆ Drill pilot holes through the back of the hangerboard and into the end of the wall stringer for two $3\frac{1}{2}$-inch No. 12 screws, then drive them in. Pull out the partially driven nails.
◆ Fasten the top of the outer stringer to the hangerboard in the same way.
◆ Screw the kick plate to the floor with four 3-inch No. 10 screws, then pull out the partially driven nails. For a concrete floor, drive in the existing nails or fasten the kick plate with a powder-actuated hammer.
◆ Toenail each stringer to the kick plate with a $2\frac{1}{2}$-inch nail *(right)*.

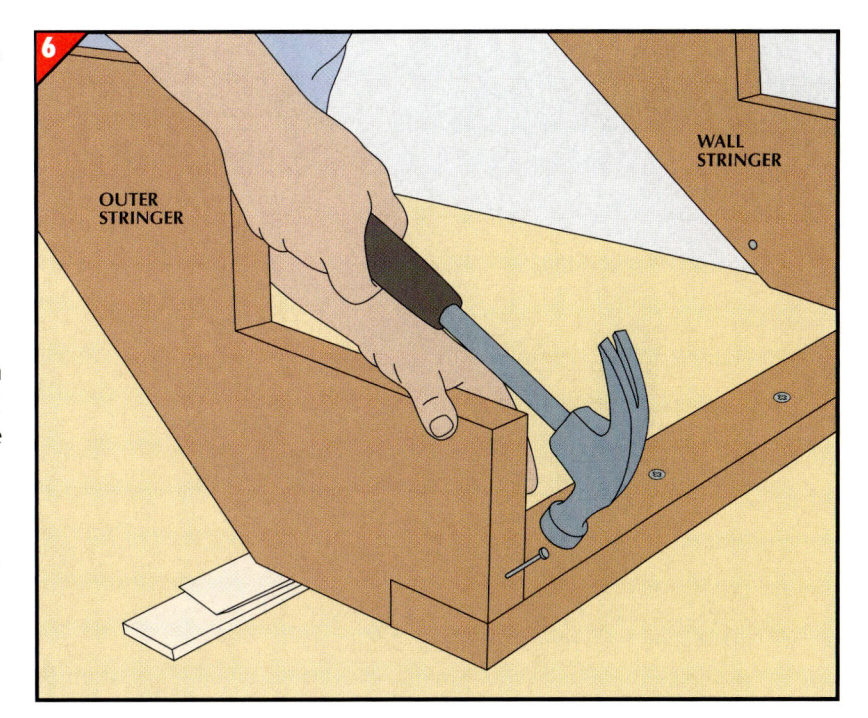

7. Adding the middle stringer.

◆ Mark a plumb line on the hangerboard $\frac{3}{4}$ inch past the halfway point between the wall and outer stringers.
◆ Position the middle stringer with its top at the level location line and its outer edge parallel to the plumb line so that it is centered between the other two stringers *(left)*.
◆ Toenail the top of the stringer to the hangerboard, driving in the nail only partway.
◆ Place a long level across the three stringers. Shim the middle stringer, if necessary, to bring it level with the other two.
◆ Fasten the stringer to the hangerboard and kick plate as you did the other two.
◆ Install a spacer, kick plate, and three stringers for the upper flight in the same way.

Cutting and Setting Newel Posts

More than any other element, newel posts establish the style of a stairway. They are also key structural elements, and serve to anchor the balusters and handrail.

Post Styles: Newels vary in appearance from simple and modern to ornate and old fashioned. A variety of turned posts are available at home centers and specialty millwork shops. Simple, inexpensive posts can be made in the shop *(below)*. If you opt for straight posts, you will not need to use curved goosenecks to attach the railing *(pages 121-122)*. Make sure your choice blends with the scale of the staircase and the style of the other woodwork in the house.

Fastening Methods: Many of the posts sold in home centers have a dowel on the bottom and are intended to be fastened to the top of the stair tread; over time, this type may work loose. Newels with extended bases that can be bolted to the stair framing are a better choice *(opposite)*. Attach these to the stair structure before the treads, risers, and skirts are installed. To add strength to the bottom post, you may want to bolt it to the floor framing *(page 111, box)*. The top newel is often a half post screwed to the wall at the top of the stairs *(page 113)*.

To hide the bolts or screws holding the posts, counterbore the holes and cover the fasteners with wood plugs.

TOOLS

Combination square
Table saw
Handsaw
Backsaw
Wood chisel
Mallet
Socket wrench
Wrench
Electric drill
Spade bit
Combination bit
Screwdriver
Carpenter's level

MATERIALS

Newel posts
Shims
Lag screws ($\frac{3}{8}$" x $3\frac{1}{2}$")
Machine bolts ($\frac{3}{8}$" x 4") and nuts
Washers
Wood screws ($3\frac{1}{2}$" No. 10)
Wood glue

SAFETY TIPS

Protect your eyes with goggles when using power tools.

A SHOP-MADE POST

Making your own posts can save money and ensure that you get exactly the look you want. As shown in the simple design at right, combining two or three elements with routed edges can create an attractive profile. First, a square of 5/4 stock is chamfered and screwed to the top of a 4-by-4 post. This is topped with a smaller square with a chamfered edge. Wood strips are then fastened to the faces of the post directly under the cap. Finally, once the posts are installed and the rails attached, the edges of the post are chamfered, leaving a square section at the top and bottom.

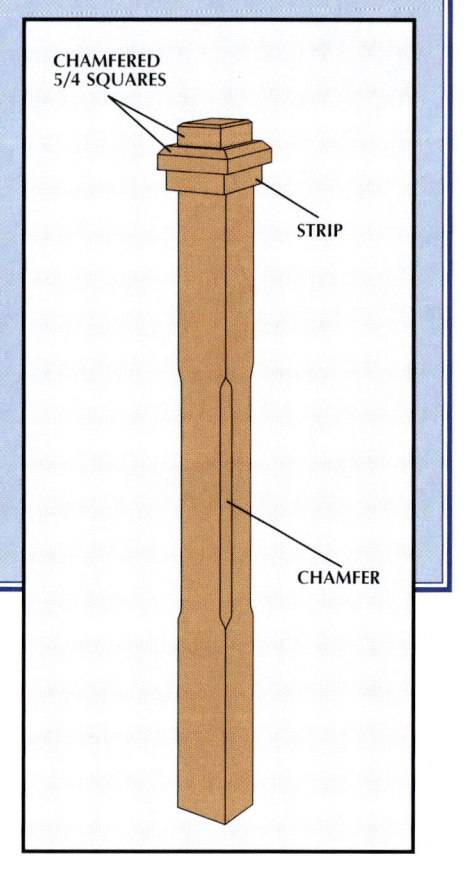

Newel bases.
For the staircase on these pages, buy three newels with square bases and cut them to the length you need. The base of the bottom post, which rests on the floor and is notched at the stringer, is cut 18 inches long. A longer one is needed if you plan to extend it to the floor framing *(page 111, box)*. The landing post, which rests on top of the lower stringer, is notched at the landing and the stringer of the upper flight; cut its base $25\frac{7}{8}$ inches long. A half-newel top post, which sits on the stringer of the upper flight, requires an $18\frac{3}{8}$-inch base. Once installed, the bases will extend above the stair treads and upstairs flooring by the same amount—$9\frac{3}{4}$ inches.

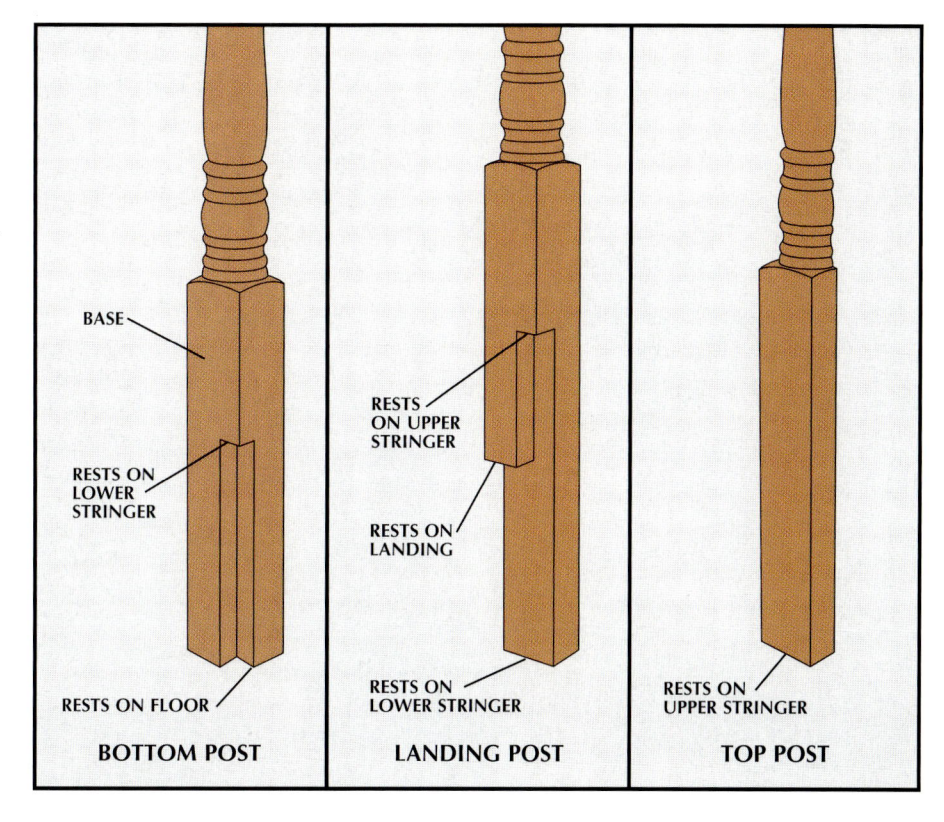

BOTTOM POST — BASE, RESTS ON LOWER STRINGER, RESTS ON FLOOR

LANDING POST — RESTS ON UPPER STRINGER, RESTS ON LANDING, RESTS ON LOWER STRINGER

TOP POST — RESTS ON UPPER STRINGER

NOTCHING THE BOTTOM MEMBER

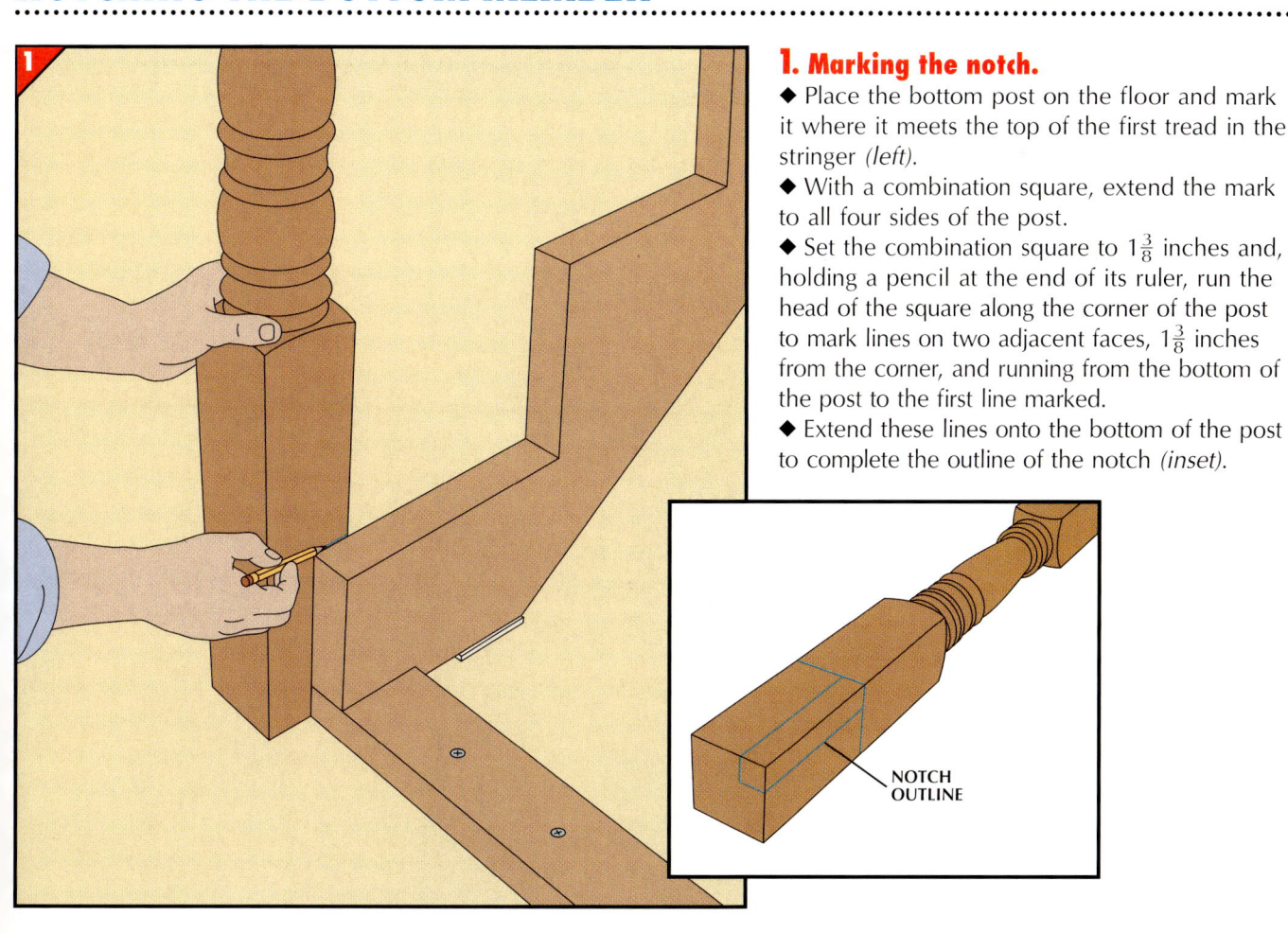

1. Marking the notch.
◆ Place the bottom post on the floor and mark it where it meets the top of the first tread in the stringer *(left)*.
◆ With a combination square, extend the mark to all four sides of the post.
◆ Set the combination square to $1\frac{3}{8}$ inches and, holding a pencil at the end of its ruler, run the head of the square along the corner of the post to mark lines on two adjacent faces, $1\frac{3}{8}$ inches from the corner, and running from the bottom of the post to the first line marked.
◆ Extend these lines onto the bottom of the post to complete the outline of the notch *(inset)*.

NOTCH OUTLINE

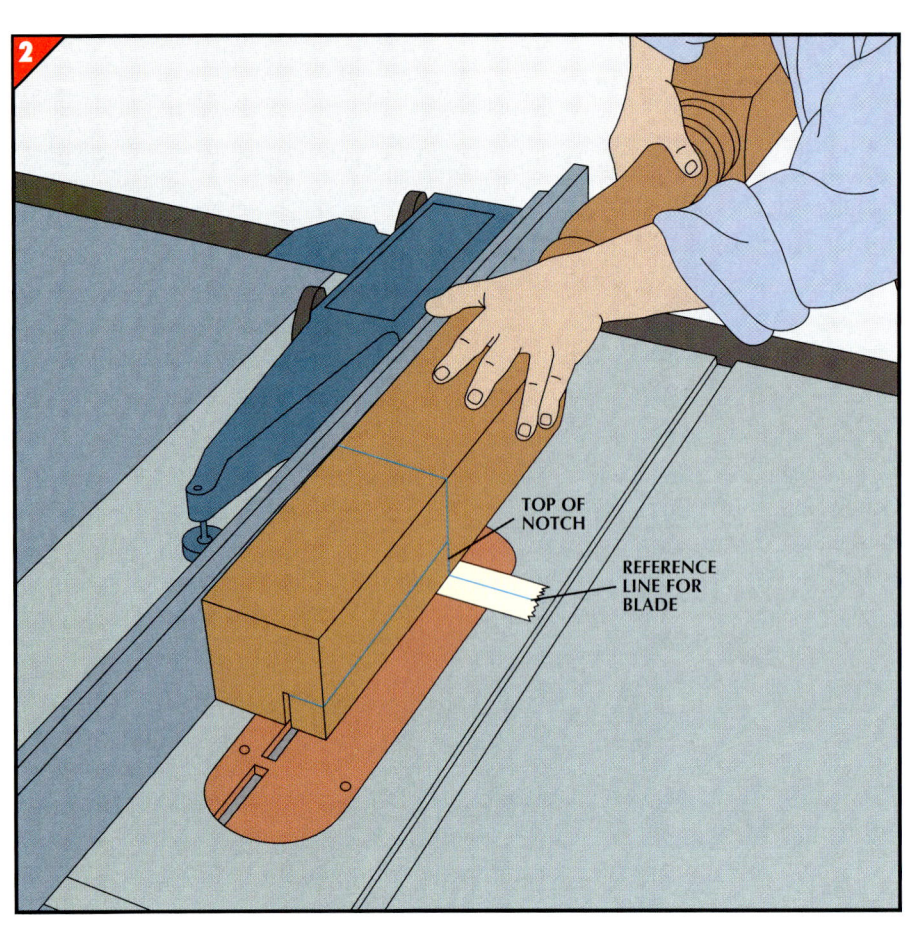

2. Beginning the notches.

◆ Raise the blade of a table saw $1\frac{3}{8}$ inches above the table.
◆ To make a reference line for stopping the cuts at the top of the notch, stick a piece of masking tape to the saw's table and mark it in line with the back of the blade.
◆ Place the post on the saw and align one of the notch marks with the blade. Set the saw's fence against the post.
◆ Start the saw and feed the post into the blade with both hands, stopping when the mark for the top of the notch is in line with the mark on the tape *(left)*.
◆ Turn the post over and adjust the fence to cut the other side of the notch, then make the cut.

3. Finishing the cuts.

◆ With a backsaw, make a diagonal cut for the top of the notch, taking care not to cut past the table-saw cuts *(right)*.
◆ Remove the waste by inserting a wood chisel between the table-saw cuts and the backsaw cuts, then tapping the chisel with a mallet. Clean up the corner with the chisel.

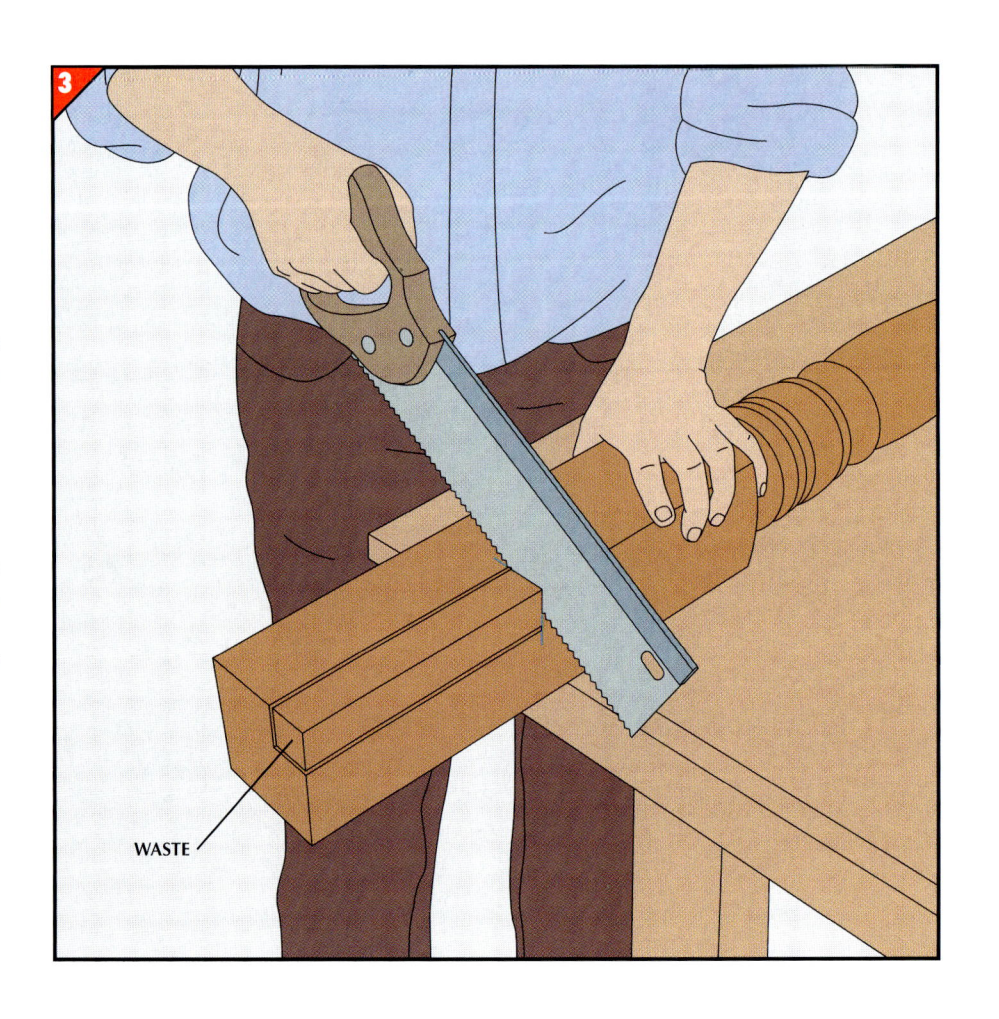

4. Installing the post.

◆ Set the post in place with its notch snug against the stringer.
◆ Starting $\frac{3}{4}$ inch from the bottom of the post, drill a counterbored hole with a spade bit for the head of a $\frac{3}{8}$- by $3\frac{1}{2}$-inch lag screw into the post. Then, with a standard bit, drill a pilot hole for the lag screw through the post and into the stringer.
◆ Slip a washer over the lag screw and drive in the screw with a socket wrench.
◆ On the adjacent side of the post, 4 inches up from the floor, drill a counterbored hole for a $\frac{3}{8}$- by 4-inch machine bolt through the post and the side of the stringer.
◆ Slip a washer over the bolt, pass the bolt through the post and stringer, add a washer, and secure it with a lock nut *(left)*.
◆ Check the post for plumb with a carpenter's level; if necessary, back off the bolts slightly and insert shims between the post and the stringer, gluing them in place.

FASTENING TO FLOOR FRAMING

The bottom newel post of a stairway takes a real beating and may work loose with time. Attaching the post to the stair framing will give it a fair bit of solidity, but securing it to the floor framing with machine bolts is the best way to lock it in place. If the finish flooring is already installed, you may not want to disturb it, but if the subfloor is exposed, consider cutting through it to access the floor framing.

Where the post happens to fall right next to or on top of a floor joist, it can be bolted directly to the joist, notching it if necessary *(below, left)*; otherwise, you will need to install blocking between the joists *(below, right)*. When the framing can be accessed from the floor below, you can install the blocking from underneath and cut an opening in the subfloor just large enough to slip the post through. Otherwise, you will need to remove a larger section of the subfloor.

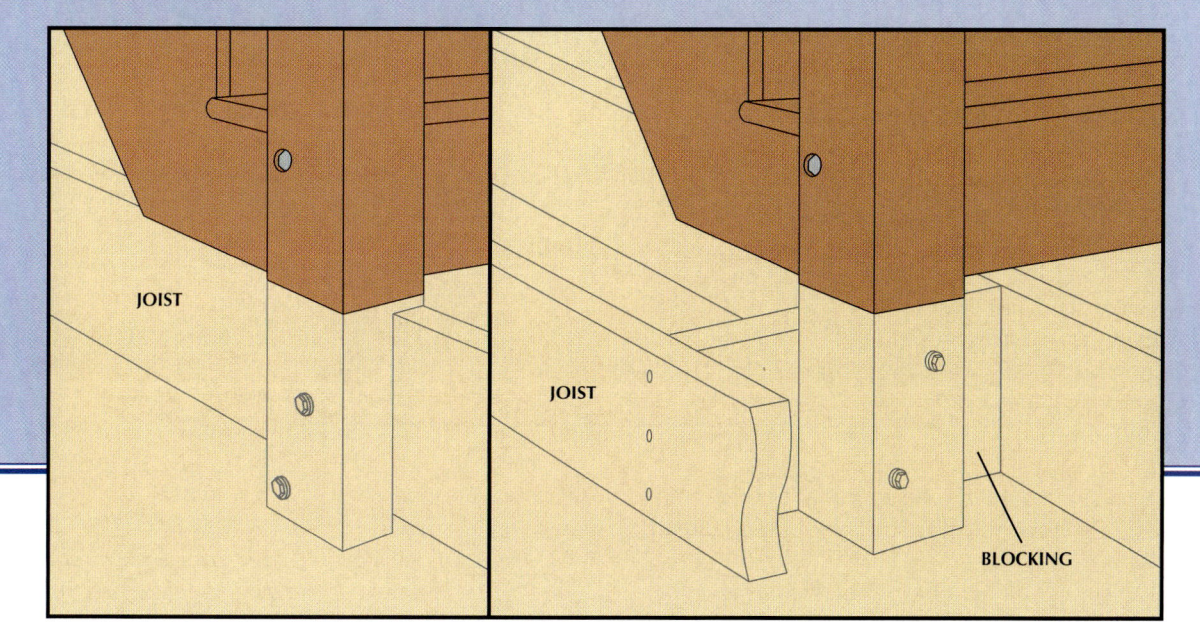

FITTING THE LANDING NEWEL

1. Cutting the first notch.
◆ Rest the landing newel on top of the stringer of the lower flight, butting it against the upper stringer. Mark the spot where the post touches the first tread in the upper stringer *(right)*.
◆ Complete the outline of the notch *(page 109, Step 1, inset)*; it will be the same as the notch in the bottom post, except longer.
◆ Cut the notch on the table saw in the same way as you did for the bottom post *(page 110, Step 2)*.

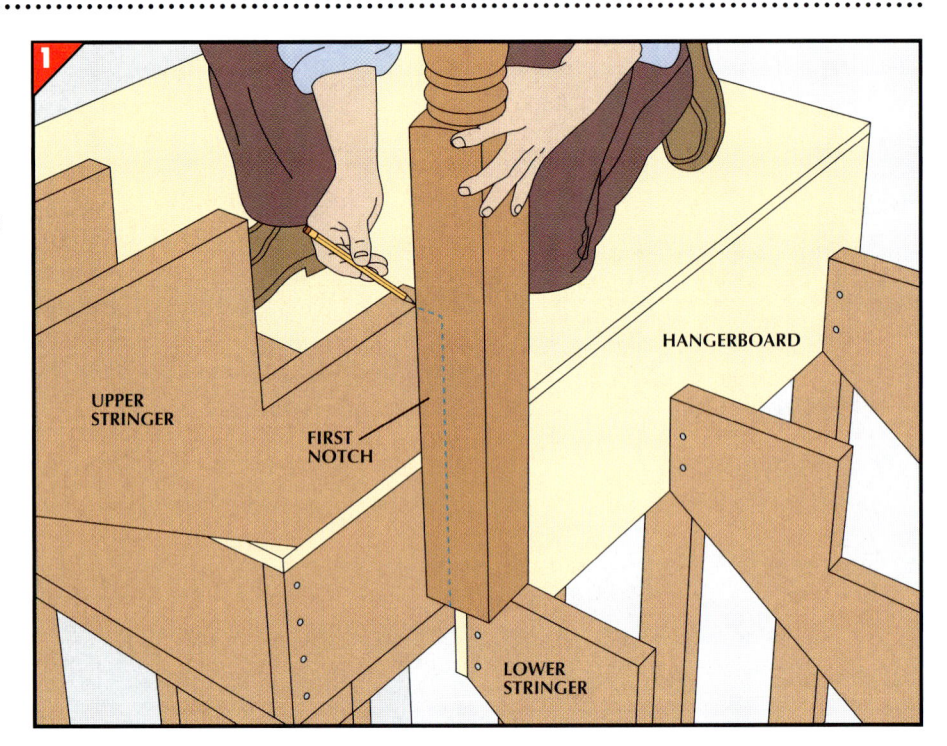

2. Making the second notch.
◆ Set the post back into the same position as in Step 1 and mark its inner face at the notch in the floor of the landing *(above, left)*.
◆ Starting at this mark, draw a horizontal line around the back of the post where it meets the landing, stopping at the first notch you cut in Step 1.
◆ Remove the post from the staircase and, on the side where you made the first mark, draw a vertical line $1\frac{2}{3}$ inches from the edge, continuing it onto the post's end *(inset)*.
◆ On the table saw, cut along the vertical line, then complete the cut with a handsaw.
◆ With a backsaw, cut along the horizontal line *(above, right)*, then clean up the cut with a chisel.

3. Attaching the post.
◆ Fit the post into place, then fasten it to the upper stringer with a counterbored lag screw and machine bolt in the same way as you did for the bottom post *(page 111, Step 4)*.
◆ Drill a counterbored pilot hole for a $\frac{3}{8}$- by $3\frac{1}{2}$-inch lag screw, centering it on the part of the post base that butts against the landing and going through the post into the landing framing.
◆ Slip a washer over the lag screw and drive it in with a socket wrench *(right)*.
◆ Check the post for plumb with a carpenter's level; if necessary, back off the fasteners and shim the post to make it plumb, gluing the shims in place.

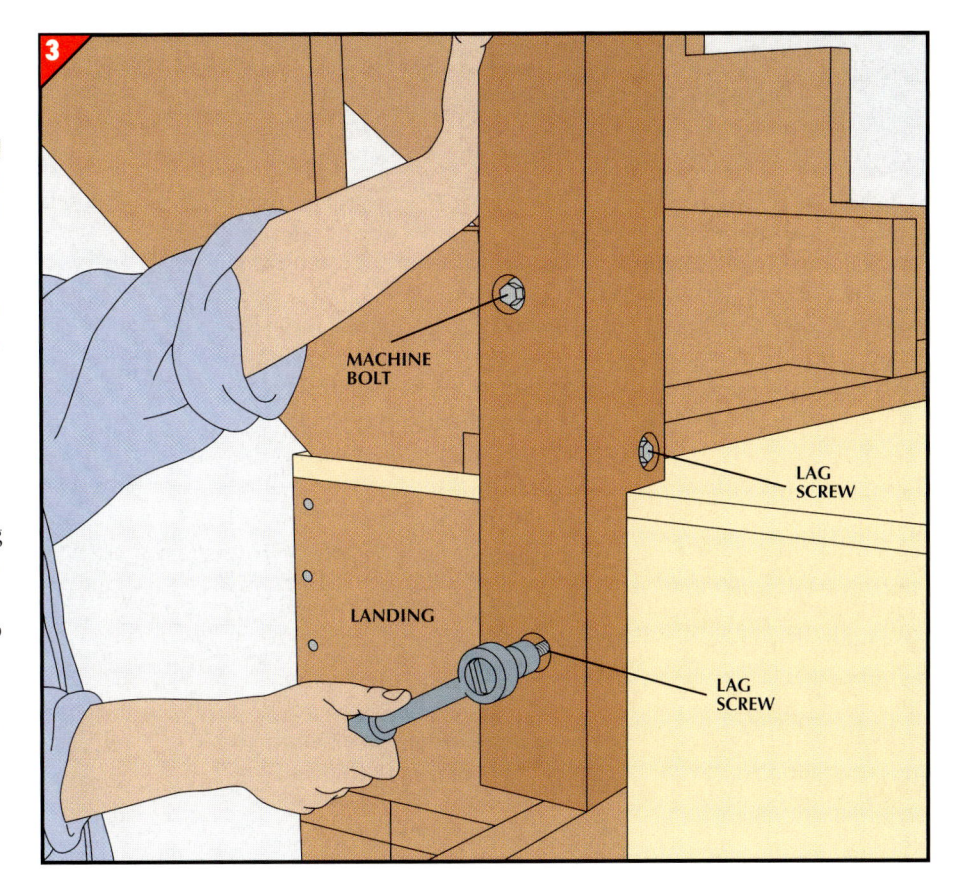

ATTACHING THE TOP POST

Fastening the post.
◆ Drill four counterbored holes for $3\frac{1}{2}$-inch No. 10 wood screws into the flat surface at the bottom and top of the half-post base.
◆ Set the post on top of the stringer and, holding it snugly against the wall, adjust it for plumb.
◆ Drill a pilot hole through each hole and into the floor or wall framing.
◆ Screw the post in place *(right)*.

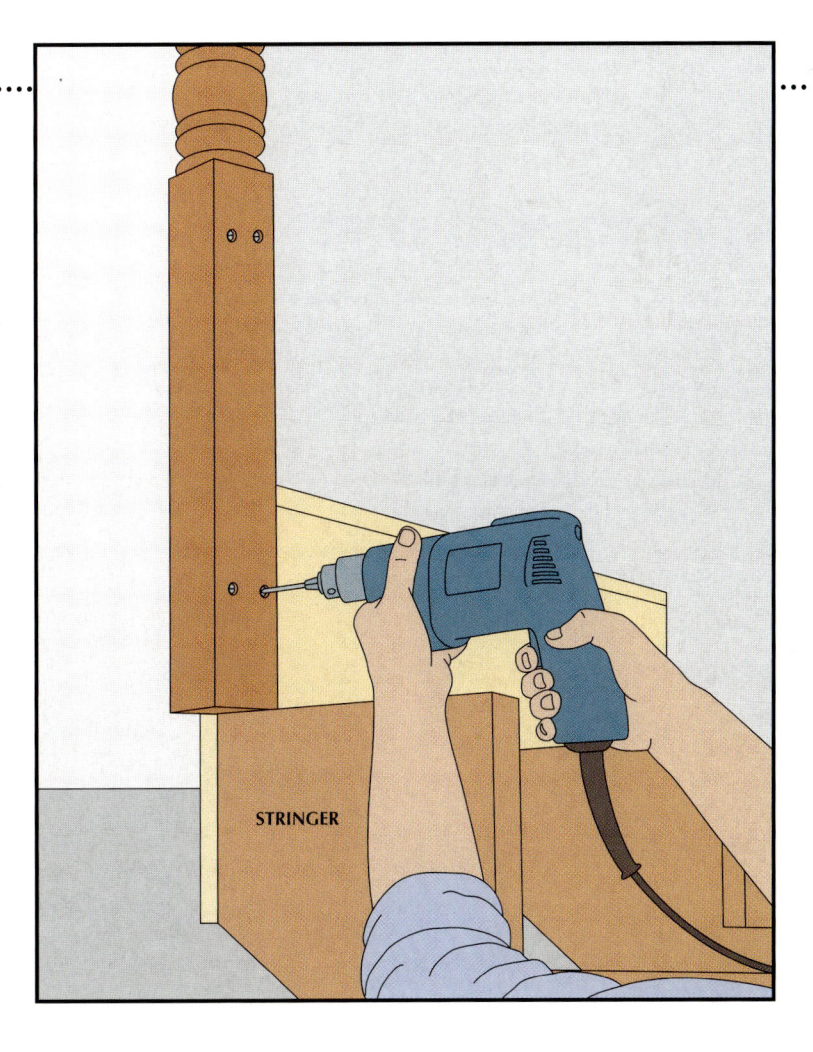

Adding Skirts, Risers, and Treads

The next step in the building process is installing the skirt boards, treads, and risers. Along with the railing, they are the most visible parts of a stairway, and require careful cutting and fitting.

Materials: If you plan to paint the staircase, you can build the parts of medium-density fiberboard (MDF). For a stained finish, purchase good-quality hardwood for all the components. To create a contrasting effect, you can use one wood for the treads and skirts, and a different wood for the risers.

Glue and glue blocks *(page 118)* will minimize squeaking when the staircase is in use.

Preparing the Pieces: Hardwood tread stock with a rounded nosing is available 42 inches long, $1\frac{1}{8}$ inches thick, and $11\frac{1}{4}$ inches wide; for stairs with a 10-inch unit run, this stock will yield $1\frac{1}{4}$-inch nosings. The risers and treads rest against the wall skirt, but on the outside ends of the stairs, the risers are flush with the outer skirt and the treads overhang it—typically by the same amount as the front nosing. Before you install the treads, round off the overhangs with a router *(page 119, Step 4)*. When preparing standard stock for treads, round off both noses and the overhangs.

Platforms for Safety: Since you will install all the risers before you put in any treads, build a couple of platforms to work on as you install the upper flight. Cut two or three sheets of plywood to the approximate size of the treads and tack them to the stringers of the upper flight. Move them up as you install risers so you always have a stable work surface.

TOOLS
Carpenter's level
Chalk line
Hammer
Adjustable T-bevel
Carpenter's square
Square gauges
Circular saw
Edge guide
Handsaw
Straightedge
Nail set
Electric drill
Combination bit
Screwdriver
Router
Round-over bit
Table saw

MATERIALS
1 x 8s
1 x 12s
Tread stock
 ($1\frac{1}{8}$" x $11\frac{1}{4}$" x 42")
Finishing nails ($2\frac{1}{2}$")
Wood glue
Wood screws
 ($1\frac{1}{2}$", 2" No. 8)
Wood putty
 or spackling
 compound
Construction
 adhesive

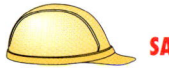

SAFETY TIPS
Protect your eyes with goggles when hammering or when using a power tool.

MAKING THE WALL SKIRT

1. Drawing a layout line.
◆ Set a carpenter's square upright on the stringer near the top of the flight *(right)*, measure along the vertical arm $12\frac{1}{2}$ inches (the height of a riser plus 5 inches), and make a mark on the wall.
◆ Use the same procedure to make a mark near the bottom of the flight.
◆ Snap a chalk line between the two marks.

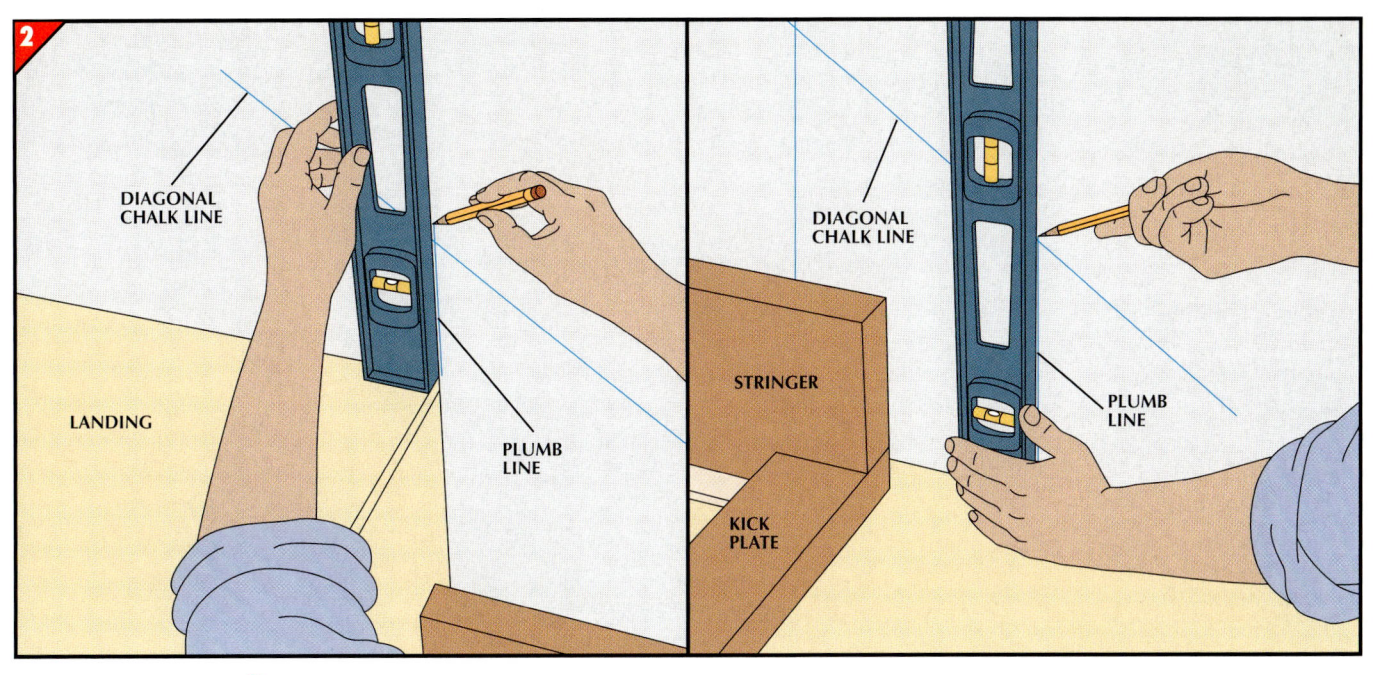

2. Adding plumb lines.

◆ Set a carpenter's level vertically on the landing, aligning one side with the front edge of the landing. Mark a plumb line that intersects the diagonal chalk line *(above, left)*.

◆ Make a mark on the wall 4 inches from the front of the kick plate. Set the level vertically on the floor, aligned with the mark, and draw a plumb line that intersects the diagonal chalk line *(above, right)*.

3. Determining the skirt length.

◆ Drive a nail partway into the wall where the diagonal line intersects the plumb line at the top of the flight. Hook a tape measure over the nail and measure to the point where the diagonal line intersects the plumb line at the bottom of the stairs *(left)*—this is the total length of the skirt.

◆ With an adjustable T-bevel, copy the angle at the top of the flight *(inset)* and transfer it to one end of a 1-by-12 about 1 foot longer than the total skirt length. Then, starting at the same end, mark the skirt length on the top edge of the board.

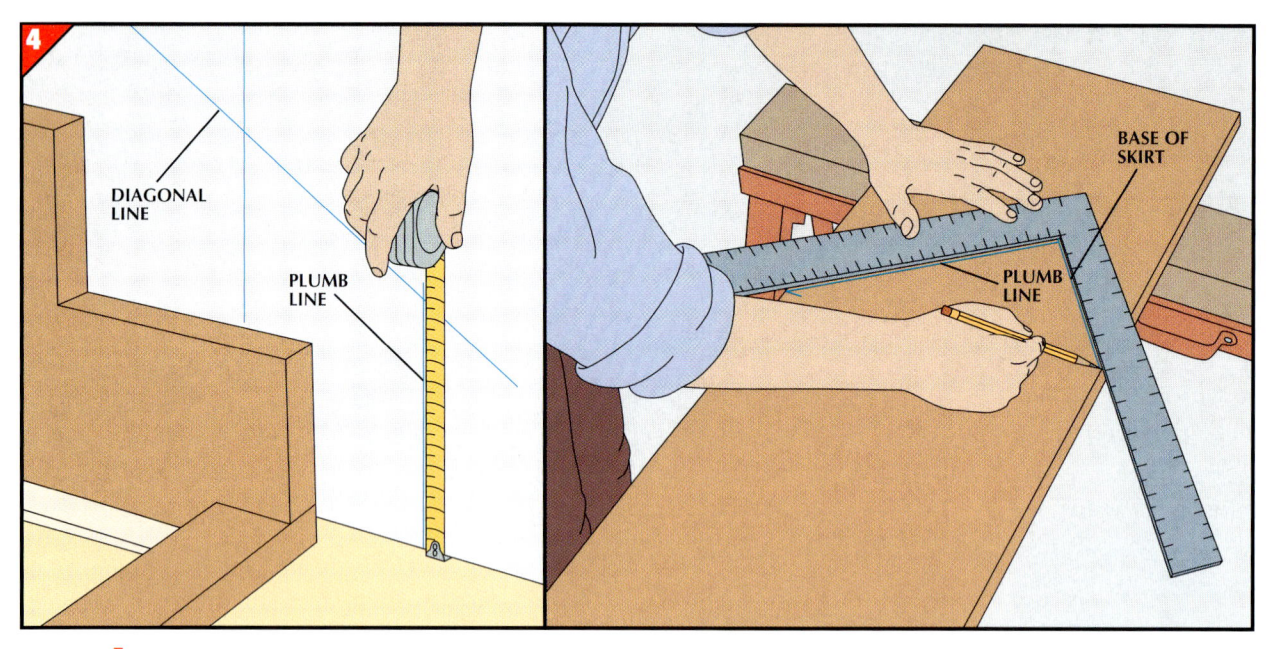

4. Marking the bottom of the skirt.
◆ Measure the plumb line on the wall *(above, left)*.
◆ With the adjustable T-bevel, copy the angle where the plumb line at the bottom of the stairs meets the diagonal line.

◆ Transfer this angle to the 1-by-12 at the point marked in Step 3, extending the line to match the length of the plumb line.
◆ With a carpenter's square, mark a line at right angles to the plumb line to square off the base of the skirt *(above, right)*.

5. Positioning the skirt.
◆ With a circular saw and edge guide, cut the skirt along the marked lines. Use it as a template to cut out a second identical skirt for the upper flight.
◆ Tack a small piece of scrap plywood as a cleat to the floor against the wall 4 inches from the foot of the wall stringer.
◆ With a straightedge, extend the stud marks *(page 105, Step 2)* a few inches past the diagonal line.
◆ Slip the skirt into place between the hangerboard and the cleat *(right)*.
◆ Nail the skirt to the wall studs with $2\frac{1}{2}$-inch finishing nails, locating the nails as close as possible to the stringer so they will be hidden by the treads and risers, then remove the cleat.
◆ Install the wall skirt for the upper flight in the same way.

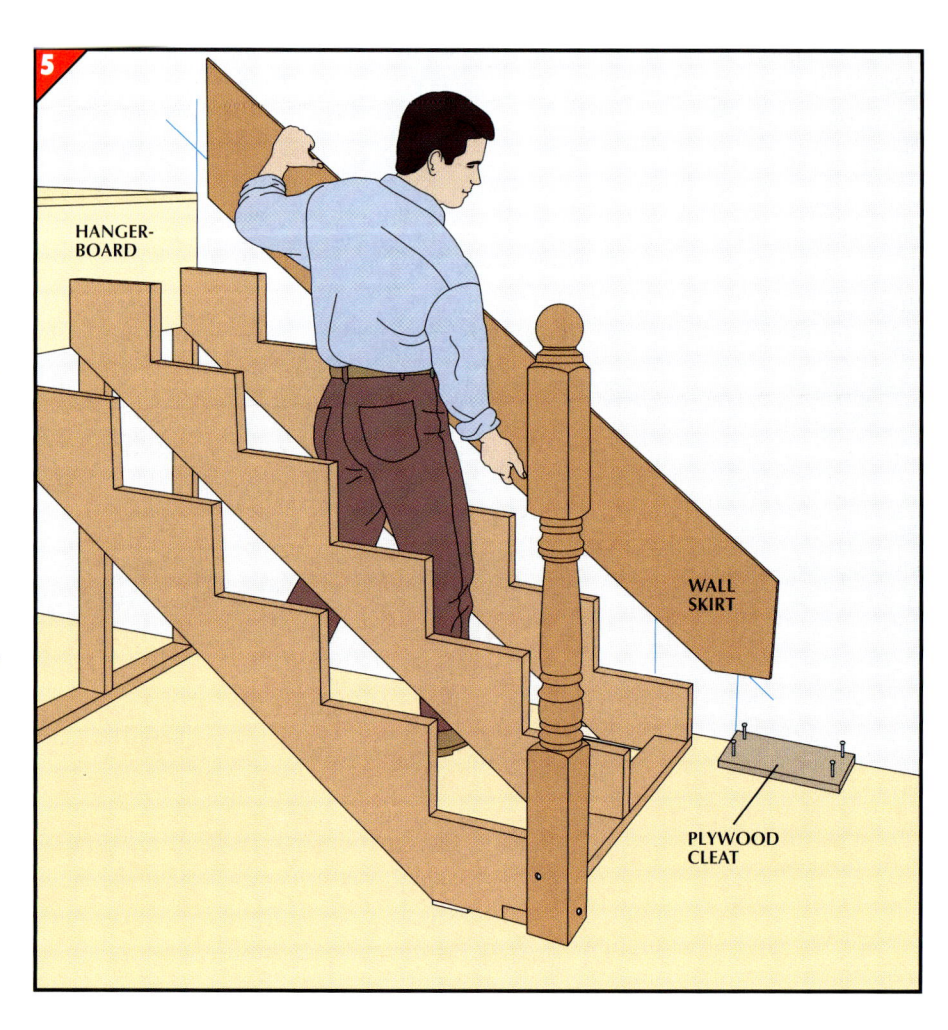

ATTACHING THE OUTER SECTIONS

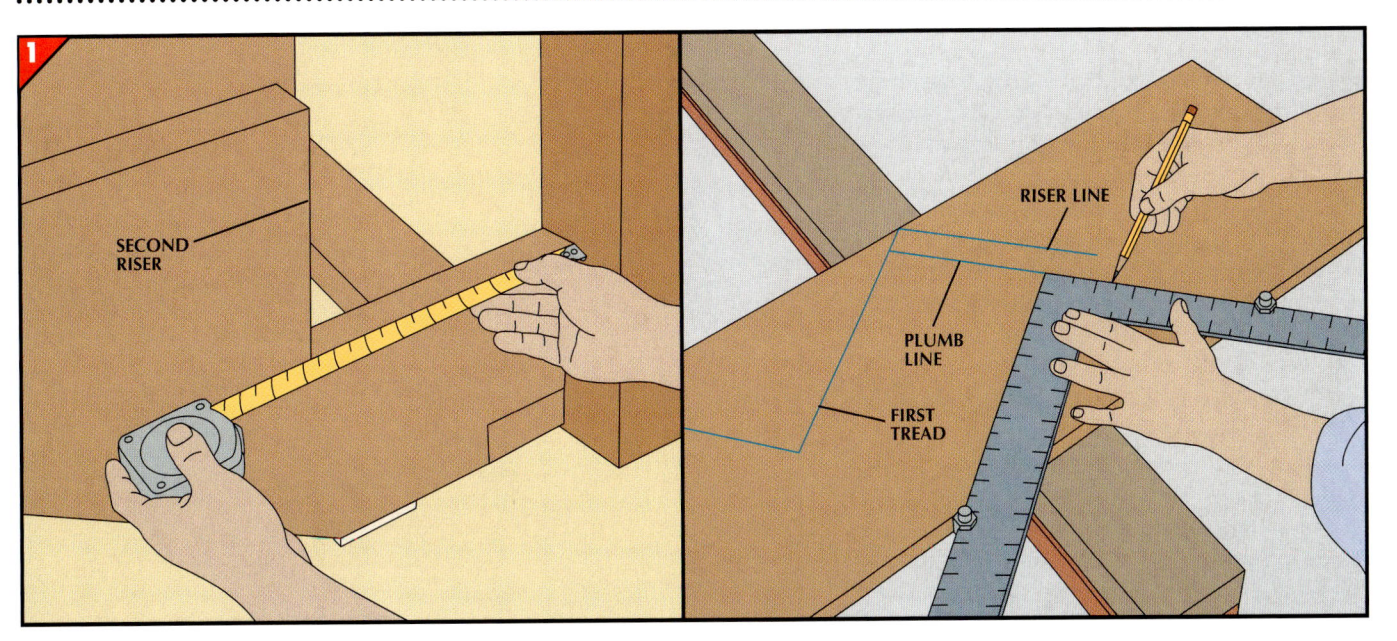

1. Cutting the skirts.
◆ Lay out a skirt board for the lower flight in the same way as for a stringer *(page 102, Step 1)*.
◆ Measure the distance from the bottom post to the second riser of the lower stringer *(above, left)*. On the skirt outline, measure this same distance along the bottom tread and make a mark. With the carpenter's square, draw a plumb line at this point (parallel to the bottom riser line). Then shift the square to the other side of the board to extend the line *(above, right)*.
◆ To square off the bottom of the skirt, measure the distance from the floor to the top of the first tread of the bottom stringer. Measure this distance along the plumb line and, with a carpenter's square, mark a line at this point at a right angle to the plumb line.
◆ Measure and mark the top of the skirt so it will fit under the landing post and butt against the landing framing.
◆ With a circular saw and an edge guide, cut the skirt along the lines, finishing the cuts with a handsaw.
◆ Measure and mark the upper skirt as you did the lower one so that it fits between the landing post and the upper wall; however, don't square off the bottom. Cut out the skirt along the lines.

2. Installing the pieces.
◆ Fit the lower skirt board into place against the outer stringer.
◆ Fasten the skirt to the stringer with a row of $2\frac{1}{2}$-inch finishing nails driven every 8 to 10 inches, 1 inch from the bottom edge *(right)*. Drive additional nails 1 inch below the top edge.
◆ Install the upper skirt in the same way so that the bottom edge runs into the landing post *(inset)*.
◆ With a nail set, sink the nails on all the skirt boards, then fill the holes with wood putty—or spackling compound, if you plan to paint the staircase.

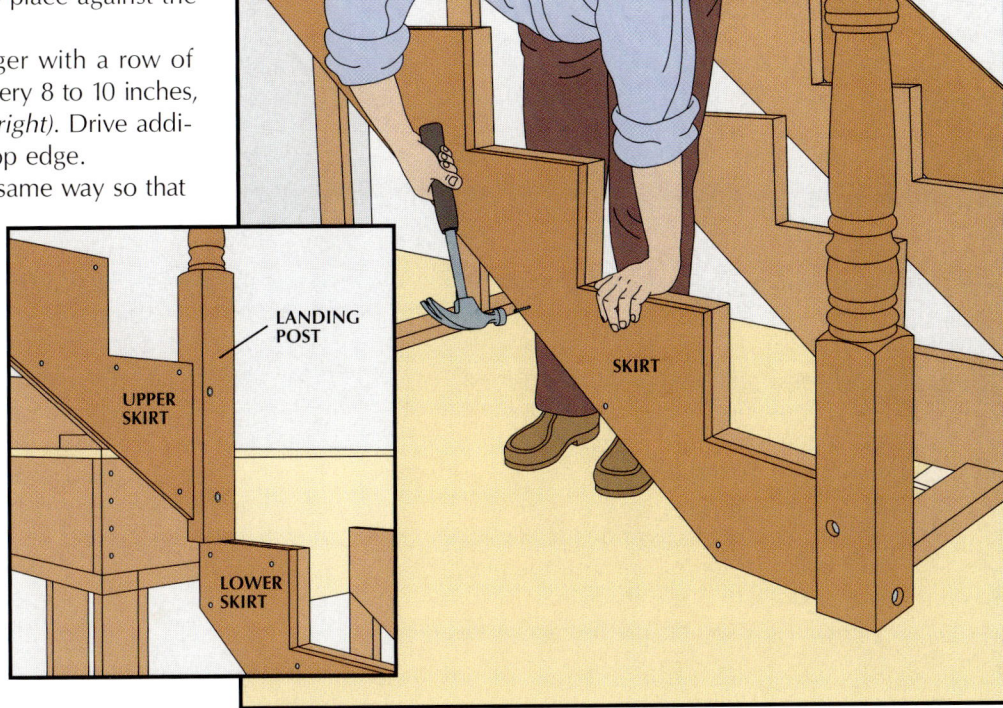

INSTALLING TREADS AND RISERS

1. Attaching the bottom riser.
◆ From 1-by-8 lumber, cut the first riser wide enough to fit the first riser section of the stringers ($7\frac{1}{8}$ inches) and long enough to fit between the wall skirt and the bottom post.
◆ From the riser stock, cut two glue blocks 2 inches long and 1 inch wide.
◆ Lay the riser face down and set the glue blocks upright on it flush with the top edge, spacing them one third of the way from each end. Drill two pilot holes for $1\frac{1}{2}$-inch No. 8 wood screws through each glue block and into the riser. Fasten the glue blocks with wood glue and screws.
◆ Apply construction adhesive to the ends of all three stringers and position the riser *(right)*.
◆ Fasten the riser to each stringer with three $2\frac{1}{2}$-inch finishing nails, angling the ones next to the post so they penetrate the stringer.

2. Fastening subsequent risers.
◆ Cut a riser as wide as the unit rise ($7\frac{1}{2}$ inches) and as long as the width of the stair structure *(page 98)*—$40\frac{3}{4}$ inches in this case.
◆ Fasten glue blocks to the riser *(Step 1)*.
◆ Apply construction adhesive to the ends of the stringers and fit the riser in place.
◆ Nail the riser to the middle and wall stringers with three $2\frac{1}{2}$-inch finishing nails.
◆ Pull or push on the outer stringer and skirt to bring the outer skirt flush with the end of the riser, then fasten the riser to the outer skirt with three finishing nails *(above)*.
◆ Install the remaining risers on the first flight, stopping at the top one.

3. Fitting the top riser.
◆ Cut the top riser as wide as the distance from the top of the stringers to the surface of the landing ($7\frac{7}{8}$ inches) and long enough to fit between the landing post and the wall skirt.
◆ Apply a bead of construction adhesive to the hangerboard in a zigzag pattern.
◆ Fit the riser in place and nail it to the hangerboard with three vertical rows of three finishing nails *(above)*.
◆ Install the risers on the upper flight as you did those of the lower flight.
◆ Sink the nails in all the risers with a nail set and fill the holes with wood putty or, if you will be painting the stairs, spackling compound.

4. Installing the first tread.

◆ For tread stock with rounded nosings, fit a router with a $\frac{1}{2}$-inch round-over bit and round the outer end of each tread. For standard tread stock, round the front noses as well.
◆ To notch a tread to fit around the bottom post, measure from the wall skirt to the post and from the riser to the post, then transfer these measurements to the tread. Make the notch on a table saw, taking care not to cut past the marks on the top face of the tread; finish the cuts with a handsaw.
◆ Apply construction adhesive to the top edge of the first riser, the three stringers, and the glue blocks, then fit the tread in place *(left)*.

5. Fastening the tread.

◆ With a combination bit, drill two counterbored pilot holes for 2-inch No. 8 wood screws through the tread into each stringer, then drive in the screws.
◆ Drill two pilot holes for screws through the back of the riser and into the back edge of the tread, then drive in the screws *(right)*.
◆ Prepare and install the remaining treads in the same way as the first, but omit the notch for the post. Notch the top tread to fit around the landing post and fasten it to the stringers—the landing framing and hangerboard will make driving screws into the back of this tread impractical.
◆ Install treads for the second flight in the same way.
◆ Install a finish floor on the landing, with a nosing to match the treads.

Putting in Handrails and Balusters

The final step in building a staircase is installing a handrail and balusters on the open side of each flight and, if necessary, a floating handrail on the wall.

Code Requirements: Check local codes to determine the required height of the handrails—usually 30 to 34 inches, as measured from the nose of the treads to the top of the railing. The distance between balusters can vary from 4 to 6 inches. A floating handrail can be added for extra safety, and is often required for stairs more than 44 inches wide. For easy-to-grasp rails, buy stock no wider than about 2 inches; set floating rails at least $1\frac{1}{2}$ inches from the wall.

Selecting Materials: Purchase rails and balusters from the same manufacturer so they match in appearance and fit together. For each tread, you'll need two lengths of baluster—a longer one for the back, and a shorter one for the front. The tops should have square ends that fit into a groove in the bottom of the handrail; but for a floating handrail, buy rail stock without a groove.

An L-shaped staircase poses a special problem for the railing design. With most commercial turned posts, the railing of the bottom flight will meet the landing newel at too low a point to attach to the flat section of the post. To solve this dilemma, you can use square posts or buy a curved gooseneck piece to link the railing to the landing post *(opposite)*.

To bring the ends of the floating handrail to the horizontal, install curved sections *(page 125);* or you can install straight end pieces instead.

TOOLS
Adjustable T-bevel
Power or manual miter saw
Handscrew clamps
Combination squares
Vise
Electric drill
Spade bit
Combination bit
Adjustable wrench
Hammer
Nail set
Screwdriver
Carpenter's square
Carpenter's level
Chalk line
Electronic stud finder

MATERIALS
Scrap wood
Handrail stock
Balusters
Gooseneck
Curved railing ends
Rail bolts, washers, nuts
Wood screws (2" No. 8)
Finishing nails ($1\frac{1}{2}$")
Handrail brackets
Wood plugs
Wood glue

SAFETY TIPS
When drilling, protect your eyes with goggles.

INSTALLING THE RAILING

1. Positioning the rail.
◆ Remove the thin wood strip, or fillet, from the groove in the underside of the rail and lay the rail on the stairs, resting it against the sides of the newels. Mark the rail where it meets the bottom and landing posts.
◆ Set a T-bevel to the angle at the bottom *(inset)* and use it to adjust the blade of a miter saw, then cut the bottom and top of the rail at that angle.
◆ Set the rail against the flat part of the bottom post and support it with a handscrew clamp; prop up the other end with scrap wood.
◆ Measure from the nose of a tread to the top of the rail and have a helper adjust the clamp until this measurement is 32 inches *(right)*.
◆ Adjust the height of the rail at the top post in the same manner and mark the post at that point. Cut the prop to hold the rail at the right height, then clamp the prop to the post.

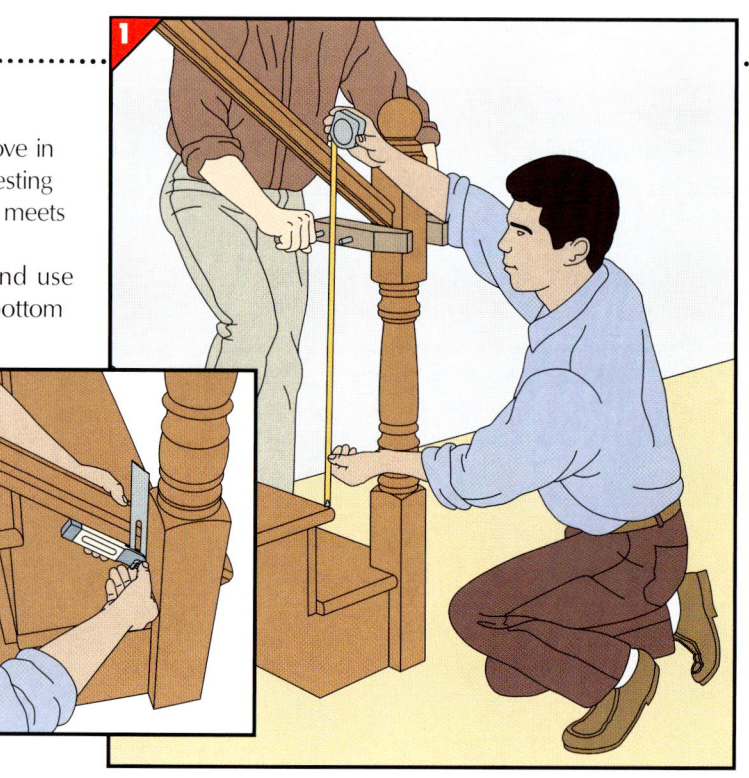

2. Aligning the gooseneck.

◆ On the top tread of the first flight, measure out 4 inches from the top riser and make a mark there. Center the dowel on the baluster over the mark, plumb it, and mark the rail at the front and back of the baluster.
◆ Remove the baluster, then have a helper set a gooseneck against the outer side of the post and level it. Adjust the gooseneck so it just crosses the handrail, with the curved section lining up with the baluster marks *(inset)*. Clamp a handscrew under the top of the gooseneck to keep it in position.
◆ Hold a combination square against the handrail, then mark the bottom of both the rail and the gooseneck at the point where they align *(right)*.
◆ Remove the handrail and make a straight cut across it—perpendicular to the bottom of the rail—at the marked line.

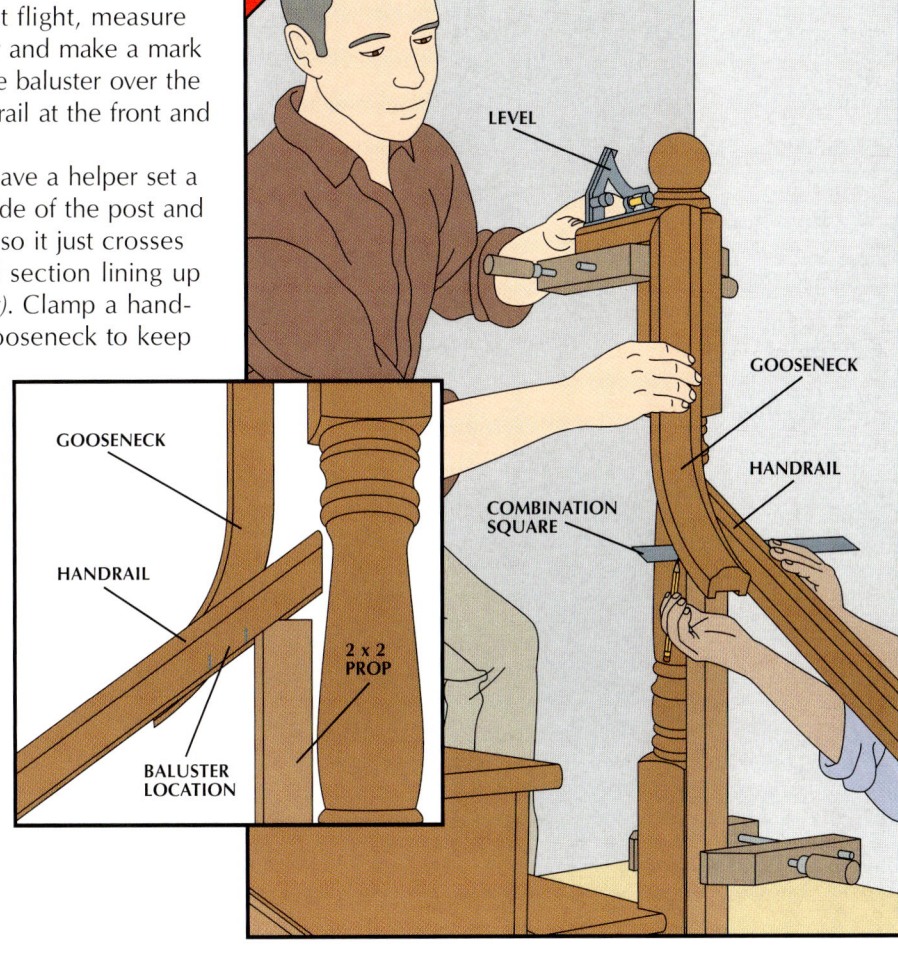

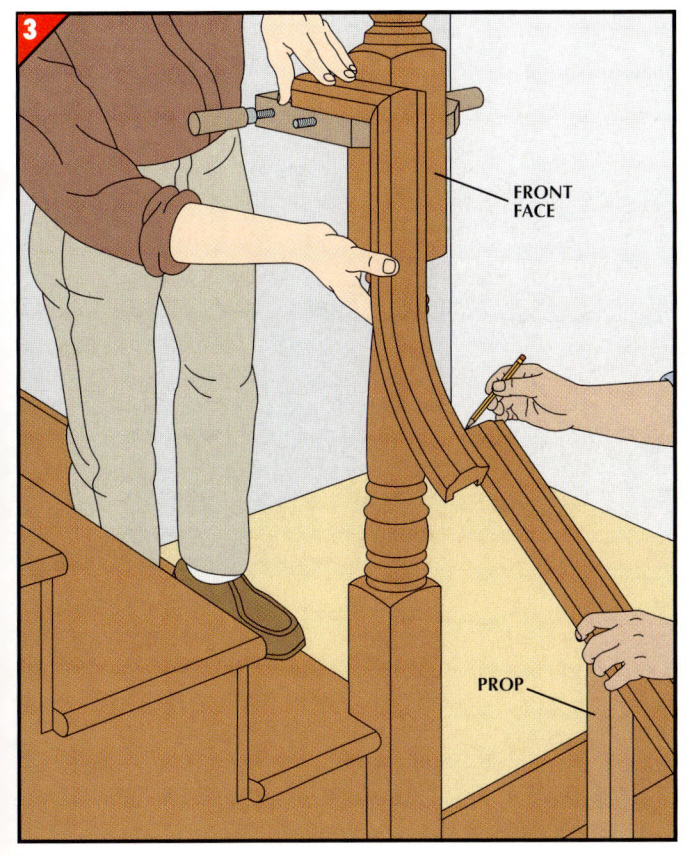

3. Marking the gooseneck.

◆ Unclamp the prop from the landing post and move it down one tread.
◆ Set the handrail on the clamp at the bottom newel and rest it on the prop, adjusting the prop until the handrail sits at the correct height.
◆ Have the helper reposition the gooseneck on the clamp and move it forward or backward until the line on the bottom of the gooseneck aligns with the end of the handrail.
◆ Make a vertical mark on the gooseneck in line with the end of the handrail *(left)*. Have your helper mark the gooseneck where it meets the front face of the post.
◆ Cut the gooseneck at the marks.

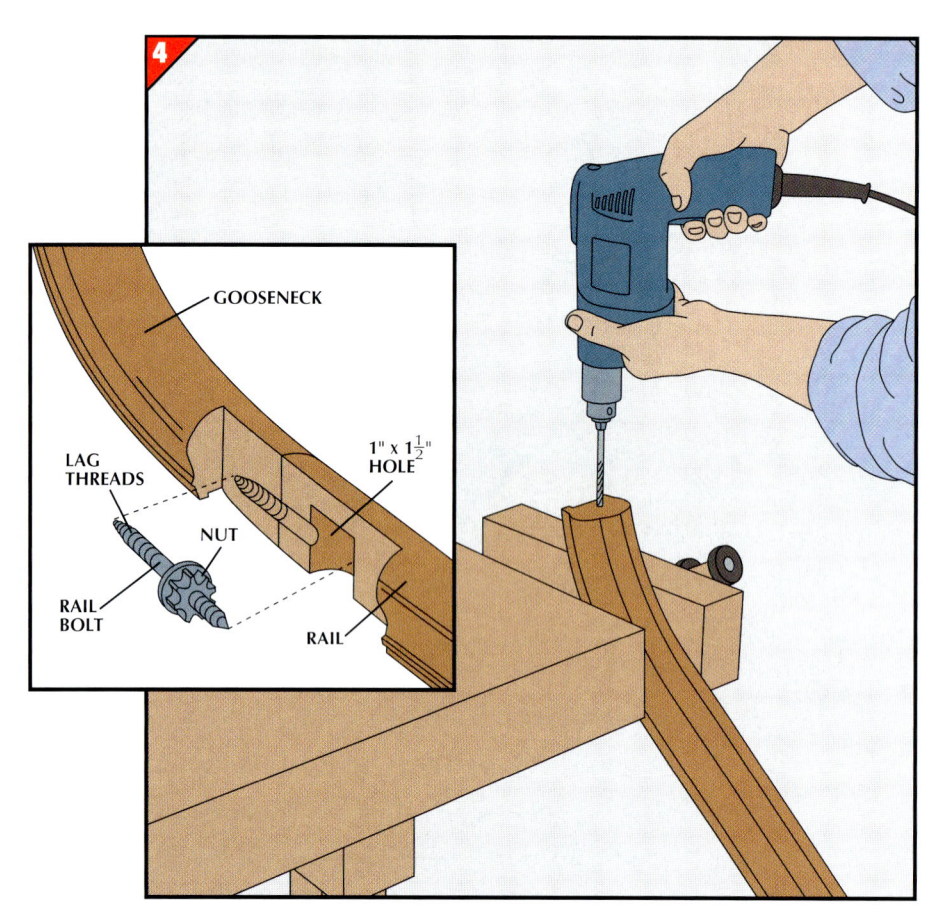

4. Joining the rail and gooseneck.

◆ Clamp the gooseneck in a vise then drill a $\frac{1}{4}$-inch hole—to match one end of a rail bolt—in the end of the gooseneck $\frac{5}{16}$ inch above the bottom of the piece and $1\frac{7}{8}$ inches deep *(left)*. Drill a $\frac{3}{8}$-inch hole for the other end of the bolt in the end of the rail.
◆ As shown in the inset, at the center of the railing bottom, drill a hole $1\frac{3}{8}$ inch from the rail end 1 inch wide and $1\frac{1}{2}$ inches deep to accommodate the nut of the rail bolt, taking care not to punch through the railing.
◆ Screw the lag-thread end of the rail bolt into the gooseneck. Work the nut as far as possible onto the bolt; turn the nut with an adjustable wrench to tighten the lag threads in the gooseneck.
◆ Remove the nut and washer, then insert the free end of the bolt into the end of the rail. Run the washer and nut onto the bolt through the hole in the rail bottom, then tighten the nut with a hammer and nail set.

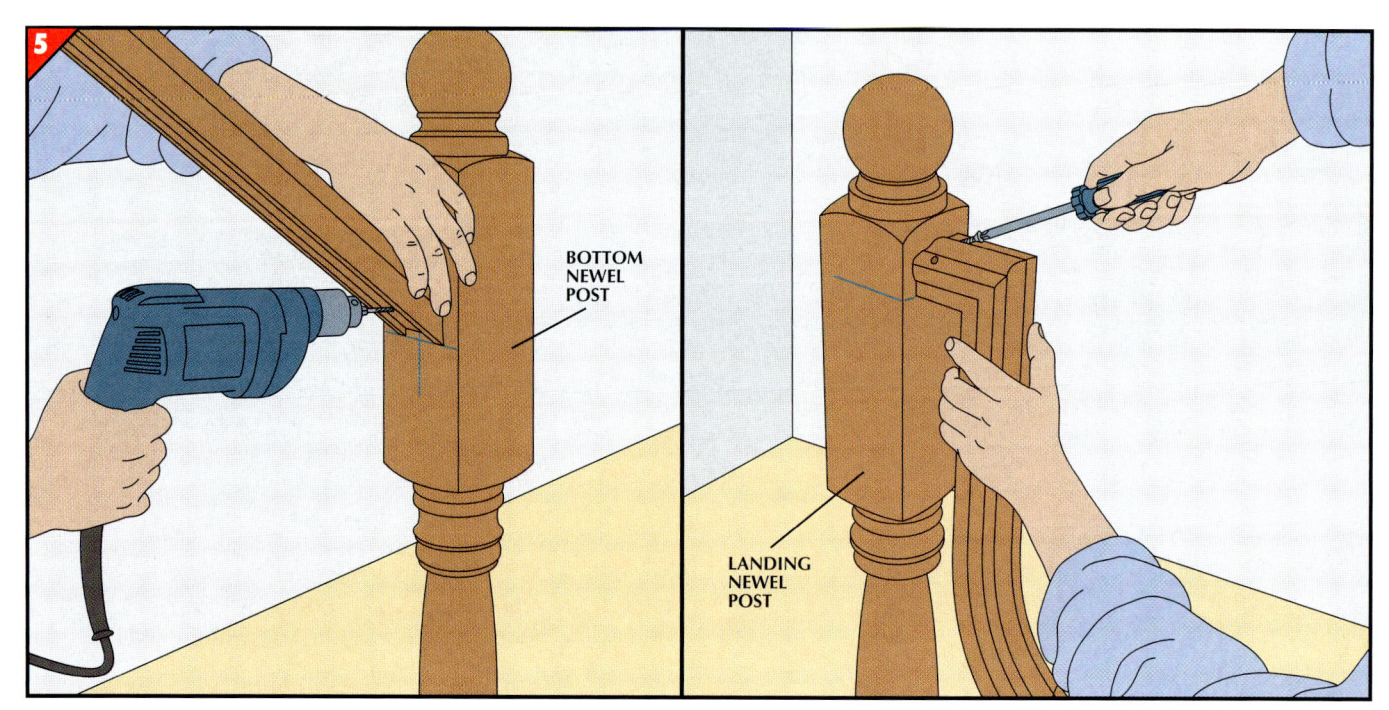

5. Installing the railing.

◆ Draw a line on the bottom and landing newels along the top of the handscrew clamps. On the landing post, extend the line onto the front face with a combination square.
◆ Set the railing in place and align the bottom with the line on the bottom newel, centering it on the face of the post. Drill two countersunk pilot holes for 2-inch No. 8 wood screws through the base of the rail and into the post *(above, left)*, then drive in the screws.
◆ Align the gooseneck with the line on the landing post. Drill two angled counterbored pilot holes for 2-inch screws through the top of the piece and into the post. Drive in the screws *(above, right)*.
◆ Fill the holes in the top of the rail by gluing in wood plugs.
◆ Install the railing for the upper flight in a similar manner, but omit the gooseneck.

FITTING THE BALUSTERS

1. Positioning the balusters.
◆ Make a mark $\frac{3}{8}$ inch from the outside end of a riser, then place a carpenter's square at the mark and draw a line on the tread *(right)*.
◆ To ensure that all the balusters along the handrail will be the same distance apart *(inset)*, first divide the unit run in half (5 inches).
◆ Set one baluster slightly back from the face of the riser below and mark where the center of the dowel in its base meets the tread.
◆ Measure 5 inches back from this position and make a second mark for the other baluster.
◆ Note the distances from each mark to the back riser (4 and 9 inches in this example) and mark the remaining treads at the same places.
◆ With a spade bit, drill a hole at each mark the same diameter and depth as the dowels on the ends of the balusters.

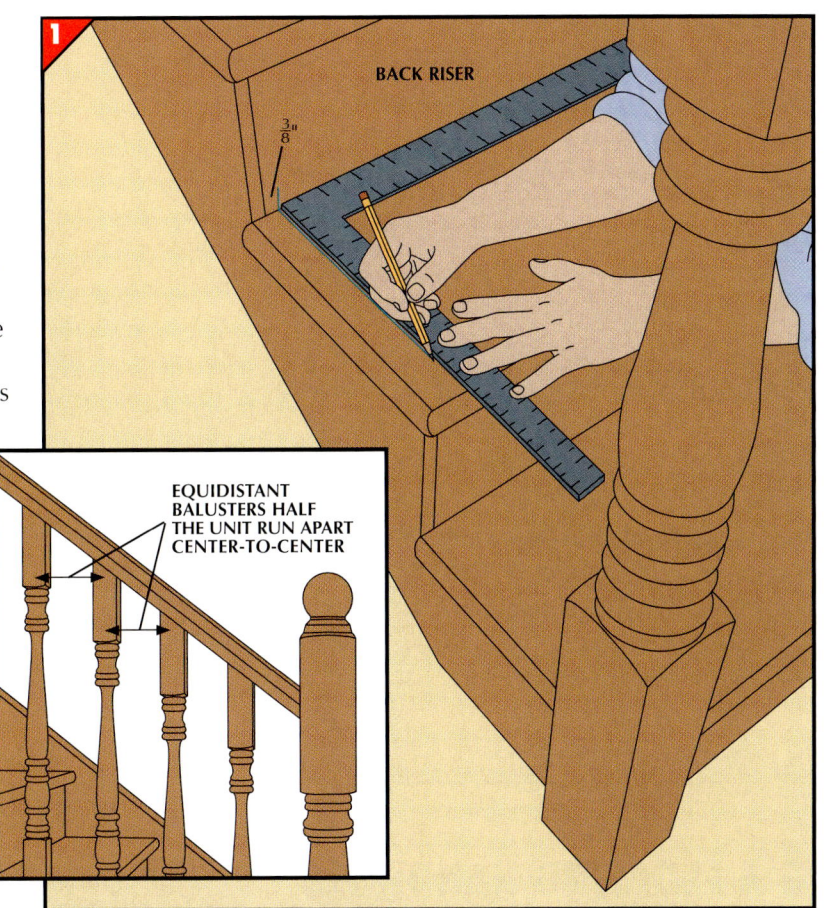

2. Cutting balusters to length.
◆ Set one of the shorter balusters into the hole at the front of a tread.
◆ Holding the baluster against the handrail, plumb it with a carpenter's level and mark a line where it meets the underside of the rail *(left)*.
◆ Remove the baluster and mark a second line parallel to the first, but higher than it by a distance equal to the depth of the groove in the bottom of the handrail.
◆ Repeat the procedure for one of the longer balusters, setting it into the hole near the back of the tread.
◆ Cut the two balusters along the marked lines, test-fit them—making any necessary adjustments—and use the pieces to mark and cut the remaining balusters, except for the one at the top of the bottom flight; mark and cut it to fit the gooseneck.

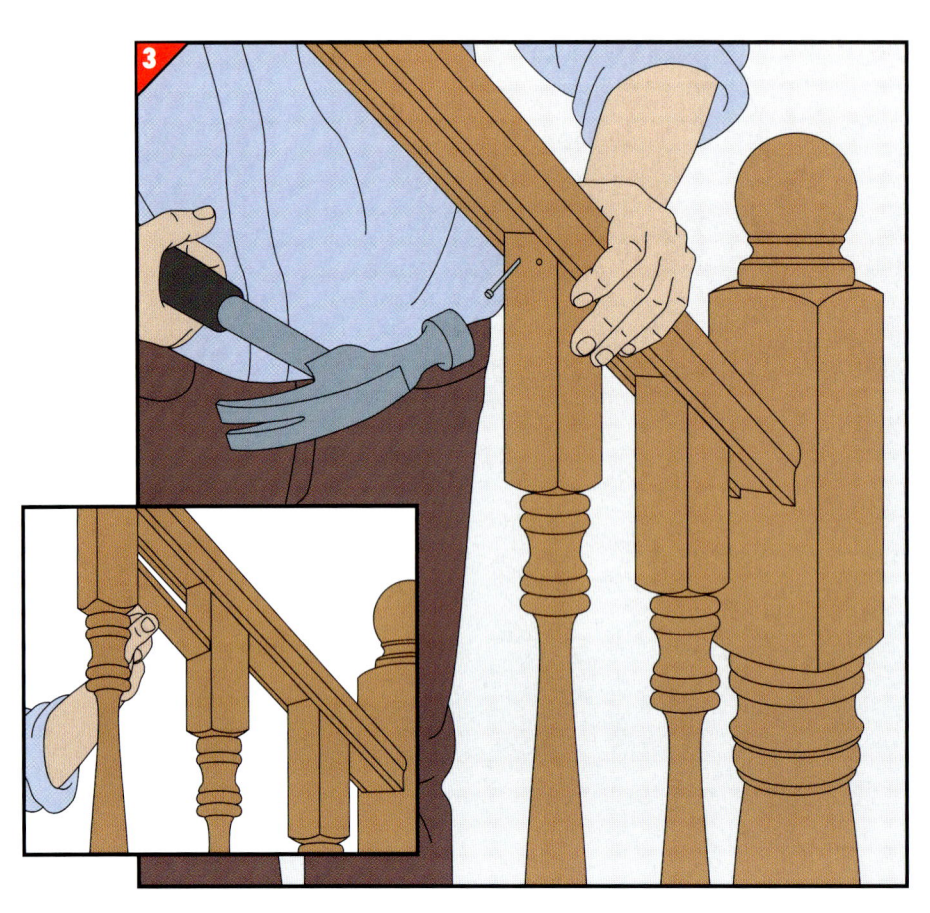

3. Fastening the uprights.

◆ Fit the baluster into place, inserting its dowel into the hole in the tread and the top into the groove in the bottom of the rail.
◆ Plumb the baluster and drill two pilot holes for 1½-inch finishing nails through it and into the underside of the railing.
◆ Remove the baluster, apply wood glue to the top, the dowel, and the hole in the tread, then reposition the piece and drive the nails to fasten it to the rail.
◆ Install the remaining balusters in the same way *(left)*.
◆ Copy the angles between the railing and the post, then cut pieces of the fillet material you removed *(page 120, Step 1)* to fit between the balusters in the groove in the rail's underside.
◆ Glue each fillet in place *(inset)* and fasten it to the railing with two 1½-inch finishing nails.

ADDING A FLOATING HANDHOLD

1. Snapping the line.
◆ Measure up from the nose of the bottom and top treads 32 inches—the height of the opposite railing—less the thickness of the railing and the height of the brackets, then make marks on the wall.
◆ With a helper, snap a chalk line between the two marks *(right)*.
◆ Trace plumb lines on the wall in line with the top and bottom of the wall skirt.
◆ Measure the chalk line from where it intersects the vertical line at the top of the stairs to where it intersects the vertical line at the bottom.
◆ Cut a length of handrail to this measured length, cutting it square at each end.

WALL SKIRT

2. Installing the brackets.

◆ Locate and mark the wall studs that cross the chalk line.
◆ Place a rail bracket over a stud mark, aligning the bottom of the bracket with the chalk line. Mark the screw holes.
◆ Remove the bracket, drill pilot holes at the marks for the screws provided, and screw the bracket to the wall.
◆ Fasten a railing bracket to each wall stud in the same way.
◆ Set the handrail on the brackets, aligning its ends with the plumb lines on the wall. Fit a bracket clip over one of the middle brackets and mark the location of the screw holes.
◆ Remove the railing and drill pilot holes for the screws provided. Replace the handrail and bracket clip, then drive in the screws *(right)*.

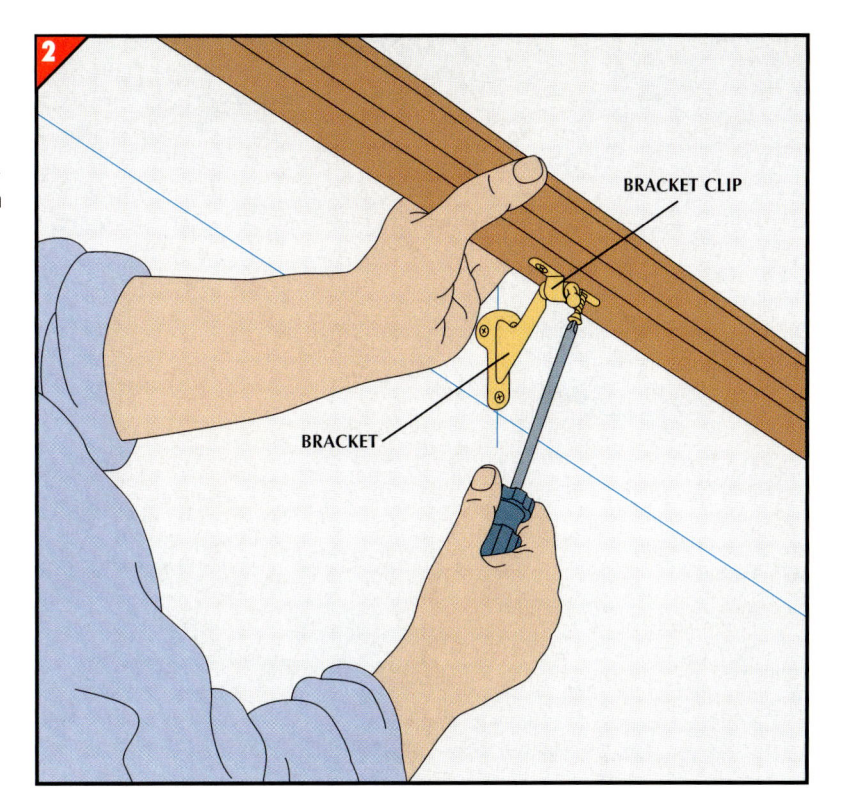

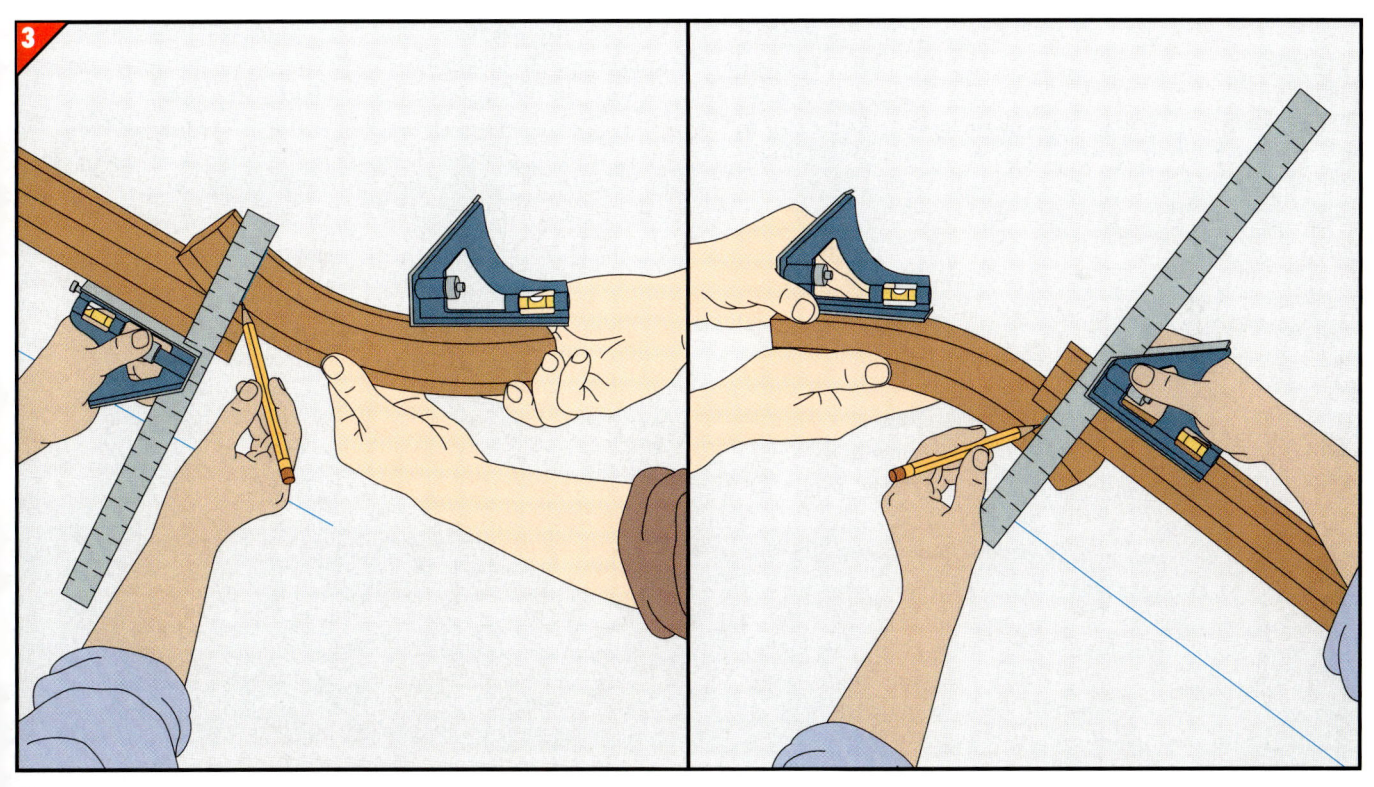

3. Fitting the curved end pieces.

◆ Rest a curved railing on the end of the handrail, adjusting its position until its end is level and its top is overlapping the handrail.
◆ At the spot where the end piece touches the handrail, set a combination square against the handrail and draw a line on the curved piece *(above, left)*.
◆ Cut the curved piece along the marked line.
◆ Mark and cut the top curve in the same way, but hold the piece under the handrail instead of on top *(above, right)*.
◆ Remove the screws holding the handrail in place and fasten the curved pieces to the ends with rail bolts *(page 122, Step 4)*.
◆ Set the handrail back in place, mark all the holes for the bracket clips, drill pilot holes, then fasten each bracket clip to the rail.

INDEX

A
Ash: 9

B
Backsaws: 110, 112
Balusters: 93, 120, 123-124
Baseboard: 12-19; cap molding, 19; corner pieces, 17; joinery, 13, 14-18; over wainscoting, 8, 26, 31; shoe molding, 8, 19; styles, 12
Beech: 9
Birch: 9
Bits: combination bits, 38, 119; drill bits, 38, 40, 74, 119; hole saws, 74; masonry bits, 40; molding bits, 11; piloted panel-raising bits, 62; round-over bits, 119; router bits, 11, 62, 119; spade bits, 74
Blades: dado heads, 60; molding-cutter-head knives, 10
Block planes: 72
Bolts, rail: 122
Butt joints: in baseboard, 13; in wainscoting, 29

C
Cap molding: for baseboard, 19; for wainscoting, 30
Carpenter's squares: 85, 102-103
Casing, door: 77-79
Casing, window: picture-frame casing, 46-52; stool-and-apron trim, 53-57
Cedar: 9
Chair rail: 8, 24-25; on fireplace surrounds, 36, 41
Cherry: 9
Circular saws: 104
Clamping: fireplace surrounds, 37, 38; frame-and-panel doors, 64
Clamps: bar clamps, 64; handscrew clamps, 38, 120, 121; pipe clamps, 37; spring clamps, 45
Combination bits: 38, 119
Combination squares: 48, 67, 78, 82
Compasses: 54
Compound cuts in crown molding: 21

Coped joints: in baseboard, 13, 15; in crown molding, 22-23
Coping saws: 15, 22
Corner pieces for baseboard: 17
Crown molding: 8, 20-23; on door surrounds, 80, 84, 86; on fireplace surrounds, 36, 41

D
Dado heads: 60
Dado joints: 65
Dimensional stability of wood: 9
Door casing: 77-79
Doorknobs: 73-76
Doors, frame-and-panel: 60-79; building, 61-64; casing, 77-79; hanging, 69-72; hinges, 69-71; jambs, 65-68; locksets, 73-76; mortises, 69, 70-71; planning, 60; quarter-round molding, 64; strike plates, 76
Door surrounds: 80-89; anatomy, 80; assembling, 81-86; installing, 87-89; jamb extensions, 87; returns, 84; siding, cutting, 88; sunburst decorations, 86
Douglas-fir: 9
Doweling jigs: 37
Dowels: 37
Drill bits: combination bits, 38, 119; hole saws, 74; masonry bits, 40; spade bits, 74
Drills: electric drills, 38; push drills, 50

E
Entryway surrounds. See Door surrounds

F
Featherboards: 10, 11, 61, 62
Finishing: trim, 8; wood species, 9
Finish nailers: 14
Fireplace surrounds: 36-41; anatomy, 36; assembling, 37-38; installing, 39-40; trim, 41
Flashing, J-channel: 88
Foam insulation: 45
Forming tools: 48
Furring strips: 27

G
Gooseneck railings: 93, 120, 121-122

H
Hammering blocks: 28
Handrails: 93, 120-122, 124-125
Handscrew clamps: 38, 120, 121
Hanging doors: 69-72
Hardness of wood: 9
Hinge-mortising jigs: 71
Hinge-mortising templates: 70
Hinges: 69-71
Hole saws: 74

I
Insulating window jambs: 45

J
Jack planes: 29
Jamb extensions: doors, 87; windows, 47, 55
Jambs, door: 65-68
J-channel flashing: 88
Jigs: doweling jigs, 37; featherboards, 10, 11, 61, 62; hammering blocks, 28; hinge-mortising jigs, 71; hinge-mortising templates, 70; miter-cutting jigs, 86; preachers, 18; reveal gauges, 48; tenoning jigs, 61; window-apron props, 57
Joints: in baseboard, 13, 14-18; butt joints in baseboard, 13; butt joints in wainscoting, 29; coped joints in baseboard, 13, 15; coped joints in crown molding, 22-23; in crown molding, 20-23; dado joints, 65; in door jambs, 65; miter joints in baseboard, 13, 14, 16-17; miter joints in crown molding, 20-22; miter joints in wainscoting, 27; miter joints in window trim, 49-52; scarf joints in baseboard, 13, 18; in wainscoting, 27, 29; in window trim, 49-52

K
Kick plates: 93, 103, 106, 107

L
Landings: 93, 96, 98-100
Locksets: 73-76

M
Mantles: 36, 40
Maple: 9
Masonry bits: 40
Miter-cutting jigs: 86
Miter joints: in baseboard, 13, 14, 16-17; correcting, 51-52; cross-nailing, 50; in crown molding, 20-22; in wainscoting, 27; in window trim, 49-52
Miter saws: 21, 52, 86
Molding. See Crown molding; Trim
Molding bits: 11
Molding cutter heads: 10
Mortises, hinge: 69, 70-71
Mullions: 60, 61

N
Nailers: 14
Newel posts: 93, 108-113
Nosing: 93, 94

O
Oak: 9

P
Paneling. See Wainscoting, frame-and-panel; Wainscoting, tongue-and-groove
Picture-frame trim: on door surrounds, 80, 89; on fireplace surrounds, 36, 41
Picture-frame window casing: 46-52
Picture rail: 8, 24
Piloted panel-raising bits: 62
Pine: 9
Pipe clamps: 37
Planers: block planes, 72; jack planes, 29; portable power planers, 72
Plaster walls and wainscoting: 27

Q

P

Plinth blocks: 77, 78
Posts, newel: 93, 108-113
Power finish nailers: 14
Power miter saws: 21, 52, 86
Power planers: 72
Preachers: 18
Push drills: 50
Push sticks: 10, 11

R

Rail, chair: 8, 24-25; on fireplace surrounds, 36, 41
Rail, picture: 8, 24
Rail bolts: 122
Railing, stairway: 93, 120-122, 124-125
Rails: door rails, 60, 61, 62, 63; wainscoting rails, 31, 34
Rasps: 48
Redwood: 9
Returns: 57, 84
Reveal gauges: 48
Reveals: around doors, 77, 78; around windows, 46, 48, 53, 56
Rise in staircases: 94, 95, 96, 102
Risers: 93, 114, 118
Router bits: molding bits, 11; piloted panel-raising bits, 62; round-over bits, 119
Routers: 11, 62, 70
Run in staircases: 94, 95, 96, 102

S

Safety precautions: mineral-spirit-soaked cloths, 33
Saws: backsaws, 110, 112; circular saws, 104; coping saws, 15, 22; power miter saws, 21, 52, 86; table saws, 10, 61, 110. See also Blades
Scarf joints in baseboard: 13, 18
Shaping trim: 8, 10-11
Shimming: door jambs, 66-67; fireplace surrounds, 39; window-jamb extensions, 47; windows, 45
Shoe molding: 8, 19
Skirts, staircase: 93, 114-117
Socket wrenches: 113
Spade bits: 74
Spring clamps: 45
Square gauges: 102

Staircases: 92-125; anatomy, 93; balusters, 93, 120, 123-124; designs, 92; dimensions, 94, 95, 96; floor openings, 97; gooseneck railings, 93, 120, 121-122; handrails, 93, 120-122, 124-125; kick plates, 93, 103, 106, 107; landings, 93, 96, 98-100; newel posts, 93, 108-113; nosing, 93, 94; rise, 94, 95, 96, 102; risers, 93, 114, 118; run, 94, 95, 96, 102; skirts, 93, 114-117; stringers, 93, 101-107; treads, 93, 114, 119
Stiles: door stiles, 60, 61, 62, 63; wainscoting stiles, 31, 33, 34
Stool-and-apron window trim: 53-57
Strike plates: 76
Stringers: 93, 101-107
Sunburst entryway decorations: 86
Surrounds. See Door surrounds; Fireplace surrounds

T

Table-saw blades: dado heads, 60; molding-cutter-head knives, 10
Table saws: 10, 61, 110
T-bevels, adjustable: 16, 21, 85, 115, 120
Tenoning jigs: 61
Treads: 93, 114, 119
Trim: buying, 8; finishing, 8, 9; shaping, 8, 10-11; types, 8
Trim, ceiling. See Crown molding
Trim, door. See Door casing; Door surrounds
Trim, fireplace. See Fireplace surrounds
Trim, picture-frame. See Picture-frame trim
Trim, wall. See Chair rail; Crown molding; Picture rail
Trim, window: picture-frame casing, 46-52; stool-and-apron trim, 53-57

W

Wainscoting, frame-and-panel: 31-35; installing, 32-35; planning, 31; simulated panels, 35

Wainscoting, tongue-and-groove: 8, 26-30; butt joints, 29; caps, 30; furring strips, 27; hammering blocks, 28; installing, 27-30; miter joints, 27; planning, 26; plaster walls, 27
Wall paneling. See Wainscoting, frame-and-panel; Wainscoting, tongue-and-groove
Walnut: 9
Window casing: picture-frame casing, 46-52; stool-and-apron trim, 53-57
Windows: 44-57; installing, 44-45; jamb extensions, 47, 55; picture-frame casing, 46-52; returns, 57; stool-and-apron trim, 53-57
Wood species, characteristics of: 9
Workability of wood: 9

127

TIME® LIFE BOOKS

Time-Life Books is a division of Time Life Inc.

TIME LIFE INC.
PRESIDENT and CEO: George Artandi

TIME-LIFE BOOKS
PRESIDENT: Stephen R. Frary
PUBLISHER/MANAGING EDITOR:
Neil Kagan

HOME REPAIR AND IMPROVEMENT:
Finish Carpentry
EDITOR: Lee Hassig
DIRECTORS OF MARKETING:
  Steven Schwartz, Wells P. Spence
Art Director: Kate McConnell
Associate Editor/Research and Writing:
  Karen Sweet
Editorial Assistant: Patricia D. Whiteford

Director of Finance: Christopher Hearing
Directors of Book Production:
  Marjann Caldwell, Patricia Pascale
Director of Operations: Betsi McGrath
Director of Photography and Research:
  John Conrad Weiser
Director of Editorial Administration:
  Barbara Levitt
Production Manager: Marlene Zack
Quality Assurance Manager: James King
Library: Louise D. Forstall

ST. REMY MULTIMEDIA INC.
President: Pierre Léveillé
Vice President, Finance: Natalie Watanabe
Managing Editor: Carolyn Jackson
Managing Art Director: Diane Denoncourt
Production Manager: Michelle Turbide

Staff for *Finish Carpentry*

Series Editors: Marc Cassini, Heather Mills
Art Director: Michel Giguère
Assistant Editor: Rebecca Smollett
Designers: Jean-Guy Doiron, Robert Labelle
Editorial Assistant: George Zikos
Coordinator: Dominique Gagné
Copy Editor: Judy Yelon
Indexer: Linda Cardella Cournoyer
Systems Director: Edward Renaud
Technical Support: Jean Sirois
Other Staff: Hélène Dion, Lorraine Doré,
  Anne-Marie Lemay, Francine Lemieux,
  Jenny Meltzer, Brian Parsons

PICTURE CREDITS
Cover: Photograph, Robert Chartier.
  Art, Maryo Proulx.

Illustrators: La Bande Créative,
  Gilles Beauchemin, Michel Blais,
  Jacques Perrault, James Thérien

Photographers: **End papers:** Glenn Moores
  and Chantal Lamarre. **10 (both):**
  Robert Chartier. **14:** Delta International
  Machinery. **17, 37, 38, 40, 48, 50, 60,
  71, 72, 86, 102:** Robert Chartier.

ACKNOWLEDGMENTS
The editors wish to thank the following individuals and institutions: Jon Arno, Troy, MI; Bois Expansion Inc., Montreal, Que.; Boulanger & Cie. Ltée., Warwick, Que.; Colonial Elegance Inc., Montreal, Que.; Delta International Machinery, Guelph, Ont.; Jon Eakes, Montreal, Que.; Louis V. Genuario, Genuario Construction Co., Inc., Alexandria, VA; Hitachi Power Tools, Norcross, GA; Tool Trend Limited, Concord, Ont.

©1998 Time-Life Books. All rights reserved. No part of this book may be reproduced in any form or by any electronic or mechanical means, including information storage and retrieval devices or systems, without prior written permission from the publisher, except that brief passages may be quoted for reviews.
First printing. Printed in U.S.A.
Published simultaneously in Canada.
School and library distribution by Time-Life Education, P.O. Box 85026, Richmond, Virginia 23285-5026.

TIME-LIFE is a trademark of Time Warner Inc. U.S.A.

Library of Congress
Cataloging-in-Publication Data
Finish carpentry / by the editors of
  Time-Life Books.
p. cm. — (Home repair and improvement)
Includes index.
ISBN 0-7835-3920-7
1. Finish carpentry—Amateurs' manuals.
I. Time-Life Books. II. Series.
TH5663.F564 1998
694'.6—dc21   98-6148